RESULTS

RESULTS

RESULTS

THE FUTURE OF PHARMACEUTICAL AND HEALTHCARE MARKETING

RJ LEWIS, SCOTT WEINTRAUB,
BRAD SITLER, JOANNE McHUGH, ROGER ZAN, STEPHEN MORALES

Published by Advantage, Charleston, South Carolina.
Member of Advantage Media Group.

ADVANTAGE is a registered trademark and the Advantage colophon is a trademark of Advantage Media Group, Inc.

Printed in the United States of America.

ISBN: 978-1-59932-507-1
LCCN: 2015939472

This publication is designed to provide accurate and authoritative information in regard to the subject matter covered. It is sold with the understanding that the publisher is not engaged in rendering legal, accounting, or other professional services. If legal advice or other expert assistance is required, the services of a competent professional person should be sought.

Advantage Media Group is proud to be a part of the Tree Neutral® program. Tree Neutral offsets the number of trees consumed in the production and printing of this book by taking proactive steps such as planting trees in direct proportion to the number of trees used to print books. To learn more about Tree Neutral, please visit **www.treeneutral.com**. To learn more about Advantage's commitment to being a responsible steward of the environment, please visit **www.advantagefamily.com/green**

Advantage Media Group is a publisher of business, self-improvement, and professional development books and online learning. We help entrepreneurs, business leaders, and professionals share their Stories, Passion, and Knowledge to help others Learn & Grow. Do you have a manuscript or book idea that you would like us to consider for publishing? Please visit **advantagefamily.com** or call **1.866.775.1696**.

There are both winners and losers in the battle against disease, but eventually, if we live long enough, we will all lose a battle with disease in the end. This book is dedicated to Mike Herman and Irene Siu and the many like them who have, who are currently, or who will eventually put up the fight of their life against a challenging disease that arrived too soon. With the help of modern healthcare and modern medicines, many of these diseases can be beaten or held in check, so we can continue living a quality life for years to come.

ACKNOWLEDGMENTS

We would like to acknowledge the many individuals cited in this book for their contributions and insights that helped shape both the direction and contents of this book. We also want to thank each of the contributing authors for lending their time, resources, and company support and the many other individuals who took the time to research, proofread, and edit along the way. We would also like to thank the families of the contributing authors who supported our team in seeing this project through to completion. Finally, we want to thank the healthcare community at large for the tireless and often thankless work you do day-in and day-out. Whether you are marketing products or using those products to improve patient lives, you are making a difference in the lives of patients who are struggling for their lives. Thank you.

TABLE OF CONTENTS

ON THE BOTTOM LINE

by Scott Weintraub and R. J. Lewis

"**S**cott, I have cancer," my best friend in the world told me on Memorial Day weekend of 2013. I think he said it once. It felt as if he kept repeating it because those three words continued to echo through my head.

"It is multiple myeloma," Mike Herman explained. He was calling me from Delaware, where he lives now. We both grew up in Marlboro, New Jersey. The bonds of friendship that began in high school have lasted more than 30 years.

Thank God he kept talking, because I had no idea what to say. I could hear him going on in detail about his condition and the nature of bone cancer and so many details, but I could not grasp much of it. His announcement had shocked me, and you can imagine how much it must have shocked him when he first got the news.

"When did you find out about this?" I finally asked. "You sound pretty knowledgeable."

"Four days ago," he said. "I am sorry it took me so long to call you. My mind felt kind of paralyzed for two or three days."

"No apology necessary," I said. "How did you learn all this about the disease?"

"Google," he said. "I just went on the Internet."

"Did your doctor tell you about it at all?"

"Yeah, but when he was telling me about it, I couldn't follow a word he said because all I heard him say was 'you have cancer.' Everything he said went in one ear and out the other. Basically, everything I'm telling you, Scott, I learned through the Internet."

As I reflected on that in the days and weeks ahead, I thought it was interesting that somebody with a life-threatening disease and access to leading oncologists would learn about his condition through the Internet. It raised all sorts of concerns and questions for me both about the quality of the information he was finding and his ability to comprehend it, but he seemed to be extremely knowledgeable. Mike is not a physician; he's not even in the medical field. How would a less educated patient learn? He needed emotional support, which his friends and family could give him, but he was on a frantic quest for information, as well. And there was no easy way to get it. The doctors and our traditional healthcare system did not give him the information he desperately needed in a manner in which he could absorb it.

—Scott Weintraub

I had a very early morning on August 13, 2014. I was meeting Scott Weintraub and a group of other friends for a day in New York City. The thirteenth anniversary of 9/11 was less than a month away, and since none of us had been to the memorial or the adjacent museum, we decided to see them both.

We got there early. After looking for the names of friends etched for eternity around the fountains, which now sit in the footprints of the original twin towers, we spent a few hours in the museum, reliving that haunting day. The visit left me in a surreal mood, almost despondent. I was still at the museum when my wife, Cathy, called at about 10 a.m.

"Irene is dead," she said, sobbing, trying to catch her breath. I sat on a bench and exhaled, swept by a wave of grief. "I'm so sorry."

Irene was Cathy's best friend. They got their first jobs together and met while in training. Later they went to business school at NYU together. Cathy was her maid of honor, and Irene, more recently, was our matron of honor.

About 18 months before she passed, Irene was diagnosed with ovarian cancer. Cathy had taken her for several of her chemotherapy treatments and had been there for her as the test results brought the painfully disappointing news.

Irene called Cathy the day she was first diagnosed with stage 3 ovarian cancer. She had just had surgery to remove an ovarian cyst and called Cathy to let her know of the diagnosis. Cathy didn't really know how to process the information or what this diagnosis meant exactly, so while on the phone with Irene, Cathy looked up stage 3 ovarian cancer on the Internet. Before Irene even had a chance to have an in-depth conversation with her doctor, and before the two hung up the phone, Cathy already knew the severity of Irene's late-stage disease and her low odds of survival.

Irene was just 40 years old when she passed away. She left a wonderful husband, who is a pastor, and a beautiful daughter and son, ages six and three. It didn't seem fair or just that she should die so young, particularly with today's advanced medicines and treatments.

—R. J. Lewis

IT'S ALL ABOUT PEOPLE

Most of us get disappointing and shocking medical news at some point, either about ourselves or about someone dear to us. As in the stories above, the diagnosis and prognosis are often alarming. Painful news leaves you in disbelief as you hope for the best. Short of an instant cure or a miracle, at these moments, you desperately seek reliable information.

We live in the information age. The Internet is awash with details on virtually any health or disease subject that you might want to research. There is enough information to keep you reading for days. However, many people are not qualified to fully process and evaluate it.

Coping with a negative diagnosis is an emotionally troubling time. Some people slip into denial, but eventually, most people want to learn as much as they can. What they need, above all, is to sincerely connect with a source of useful and relevant information.

These moments represent incredible opportunities for the pharmaceutical and healthcare industry to help and be of service.

Because today's healthcare system swims in a sea of data and separates the patient who needs the product from the physician who prescribes it and from the insurer who reimburses it, it is easy to forget, or never really understand, the individual patient's personal story.

Each of our customers is a living, breathing person. All our customers have families, hopes, and dreams that are being disrupted by illness. When the pharmaceutical marketers of the future communicate with patients, those communications will need to be personal and meaningful, as well as useful and valuable if we expect them to take actions that yield results.

Healthcare marketing is a powerful tool that is about more than profits: It can deliver the information that will add years to people's lives. To gain market share and grow the brand is extremely important to any business, of course. But any business is all about people. As an industry, we need to grasp the opportunity to serve patients more intimately and immediately.

In doing so, the industry is remaining relevant in this new era, and relevance shows itself powerfully on the bottom line. To service customers in a more personal way, the pharmaceutical marketer of the future will focus on adding significant value beyond the medication itself.

We are in a noble and inspiring business. The essence of our industry is to help people get better, manage their conditions, and extend their lives. What an awesome reason to get up every day and go to work!

HUGE OPPORTUNITY, HUGE CHALLENGE

We live in an age when relevant information can be readily found. The world is awash in more information than ever before. The ability to deliver targeted quality information, education, and services when they are needed most represents a massive opportunity for our entire healthcare ecosystem.

It is an opportunity to help improve patient outcomes by engaging existing customers and new customers resulting from healthcare reform measures, while also growing our business.

In years past, pharmaceutical companies would come out with products that were significantly better than what was already on the market. Physicians and patients flocked to use those products. More recently, approved products for major diseases have only been marginally better than existing therapies, and most products are also competing with generics.

Today, you can get a product for $5 that would rate, say, an 80 on an effectiveness scale of 1 to 100. You can get another product that would rate just a few ticks higher on that scale, but instead of five dollars, it costs $50 or more.

We have a lot of government regulation, with more to come. New drugs must be shown to be significantly better, and larger and stronger managed-care companies are changing copays and widening the price gap as generics get cheaper and branded drugs get more expensive. Anyone can market a product that is significantly better than the competition's. It is another story if the product is only slightly better. To win in today's world requires a better approach.

What will constitute great pharmaceutical marketing of the future? To be effective, it must lean into a new outcomes-based paradigm and deliver results better than the competition. It will involve figuring out where in the country a product will work better, in terms of both efficacy and revenue. It will involve more than sending out an army of sales reps who often can't get in the door anymore. It will take advantage of new digital promotion tools and targeting. It will harness big data for a far better understanding of competitive advantages, ideal patient populations, and how medicines can be personalized to the individual.

Superior pharmaceutical marketing of the future will involve understanding the nuances of why certain patient populations excel with a product and why others struggle. This will come about through an understanding of real-world data, the only data that really matters. As systems of care continue to evolve and government takes an active role, figuring out how to win as a marketer becomes ever more complex.

This book is for all the players in the pharmaceutical industry: the developers, manufacturers, and marketers, whose business is the products that our society increasingly needs amid so many health challenges. Those players include pharmaceutical, life sciences, biotech, and medical device companies as well as the agencies, partners, publishers, suppliers, and the many service providers that support them in their marketing efforts.

Each of the chapters in this book contains specific information on trends in healthcare that you need to understand if you are to better market your products.

The perfect storm has been brewing for the pharmaceutical and healthcare industries, and we must be ready at the helm for what lies ahead.

A LOOK AT THE CHAPTERS TO FOLLOW

In this book, several of the keenest minds in the industry will be sharing their views and ideas about where this industry is heading. The table of contents outlines the key categories, or areas of leverage, that the authors believe will drive pharmaceutical and healthcare marketing of the future. We believe pharmaceutical marketers must develop a deep understanding of each of these disciplines to remain relevant and profitable.

Each chapter examines an evolving trend. Here is a synopsis of what you will learn in the pages ahead:

Chapter 1: Regional Marketing

In the marketing of the future, one size doesn't fit all. That is an approach that no longer works. Healthcare marketing is becoming more local. Many marketers recognize the fact that different parts of the country have different needs but are unsure of the best strategy to address this challenge.

Today the typical brand has a 700 percent variation in sales among markets. For example, in New York City, 5 out of every 100 prescriptions for high blood pressure might be for one specific drug. In Dallas, it might be 35 out of 100.

Clearly, marketing efforts need to be tailored to the region. But how? The answer lies in the six Ps of pharmaceutical marketing: product, place, patient population, payer, provider, and prescriber. Each can vary significantly in local markets, greatly influencing product performance.

Let's look briefly at those six Ps. The product differentiation, efficacy, and competition may differ. The sales reps in one place might speak to doctors more often than the reps in another place—perhaps once a week rather than once a month. The population varies between old and young, rich and poor, and the racial and ethnic demographics vary, all of which can affect both the sales and effectiveness of a product. In addition, the payers, a combination of the US government and various insurers, could be charging patients varying amounts for the same product. And one locality's provider system—that is, the hospitals—may favor one product or procedure over another for whatever reason. And finally, how many prescribers of that product are there in the region? One region, for example, might have many specialists writing scripts for a niche drug, while another region's prescribers might be mostly family doctors.

Within each of those variables are dozens of others, and it is essential that industry marketers have a structured process to analyze and understand them.

One size no longer fits all in healthcare marketing. At its core, healthcare is a local business. Different needs in different regions significantly influence sales. How can a pharmaceutical company sort through hundreds of variables for insights into how to reach those regions effectively and efficiently?

Chapter 2: Digital Marketing

As both the media and the attention spans of your audience become more fragmented, the delivery of well-timed, personalized, and informative messages becomes increasingly important. Marketers must be heard above the noise, deliver meaningful value, and build real brand loyalty.

Customer's expectations are changing. Both patients and prescribers expect more value and service from pharmaceutical companies. While concerns over privacy are real, they are, nevertheless, eroding as customers can receive free, quality information in exchange for their data. Good marketers will respect their customers' desire for privacy while also capitalizing on the data those customers willingly provide by leveraging targeting and personalization technologies that deliver the right message to the right person at the right time.

Marketers of the future will track whether every message was delivered, how often, and for how long it was viewable on a screen, and they can appropriately vary, optimize, personalize, and report on each of their messaging engagements throughout the full continuum of care.

New digital channels, such as wearable and ingestible technologies, will contribute to patient feedback and support. One-size-fits-all messaging, delivered via mass media, is simply no longer relevant and only contributes to the noise.

Chapter 3: Promise of Big Data

So much more information is available than ever was imagined in the past. The essential question is this: How can we make use of that avalanche of information for better marketing?

The challenge in this age of big data is how to rally the expanding databases for potentially huge benefits. How can we sift through those mountains of data for improved sales? If we can make sense of it all, we can learn to take the right marketing actions.

For example, patients with the stomach bacterium H. pylori often get a false diagnosis of GERD. The doctor might prescribe a proton pump inhibitor (PPI) to no avail, as many doctors do. By harnessing big data, a marketer can learn whether another condition is present in a cohort of patients in addition to H. pylori. That way, the marketer is not just targeting physicians who see lots of patients with H. pylori. The marketing becomes this: "We want to market to doctors who write lots of scripts for proton pump inhibitors and whose patients likely have this particular comorbidity."

Those are the physicians who need to be educated about H. pylori because they very often misdiagnose GERD. We can look at the data on the patient level, aggregate it up to the physician level, and learn that Dr. Jones may be misdiagnosing lots of patients while Dr. Smith is, likely, more accurate. We can appropriately personalize the education of each physician.

That is the promise that we are seeing in big data. We can pinpoint where everyone needs to learn and improve.

Chapter 4: Evolving Systems of Care

Healthcare reform—driven largely by the Affordable Care Act—is creating a number of changes in our healthcare system as well as an influx of millions of patients, creating both opportunities and new challenges for manufacturers.

We are seeing a fundamental change in expectations for patient care and subsequently how healthcare providers are compensated. Payers (employers and health plans) are trying to shift the risk of patient management to providers. Providers will be managing the comprehensive care of a patient, and therefore their pay will increasingly be based on their performance and the quality of their results in terms of patient health outcomes.

The cost implications of care are increasingly visible to all key stakeholders in the healthcare continuum, including patients and caregivers. Once, a great many physicians were sole proprietors. Now, through these evolving systems of care, more than half of the physicians in the United States work for large hospital systems or large group practices. These are run by business people whose goal is to improve the care and lives of their patients but who also need to focus on financial vitality of their practices or institutions. Payers are also looking to deliver the best care in the most cost-effective manner. For patients, the cost burden of healthcare is being increasingly shifted to them through higher copays, co-insurance, and deductibles.

With these forces in mind, care may go something like this: When patients come in with high blood pressure, they try a generic drug for six months as initial therapy. If that doesn't work, an alternative generic therapy may be prescribed, and if further therapy is needed, another generic option may be pursued before the more expensive "branded" therapy is considered. When cost consideration is highly important, that makes sense. Newer therapies are highly scrutinized and evaluated to ensure they enhance or deliver a higher standard of care.

But let's say you are a physician, and your wife is the patient. You want the most effective medicine for her, whether it is the least expensive or not. What you would prefer, however, is not necessarily what the system advocates. Physicians are now encouraged, and in fact incentivized, to follow treatment protocols that are based on the evaluation of the available evidence on the available therapies, patient outcomes and cost to the system.

Pharmaceutical manufacturers have to understand this evolving market and determine how to work best with these big systems of care, given how the incentives are aligned, protocol, and new therapy evaluation requirements. Under the Affordable Care Act, outcomes (combining both clinical and economic) have become incredibly important, influencing how care is provided and how manufacturers should develop and market their products and services.

The Startling Truth

Our purpose in this book is to educate pharmaceutical marketers about the startling truth that unless they continue to develop expertise in this rapidly changing field, they risk becoming irrelevant.

CHAPTER 1

REGIONAL MARKETING

by Scott Weintraub

THE INDUSTRY'S MASTER PLAN HAS LONG BEEN TO TRY TO MOTIVATE THOUSANDS OF SALES REPS TO TALK TO AS MANY PHYSICIANS AS POSSIBLE, HAMMERING HOME WHY A CERTAIN DRUG IS BETTER THAN THAT OF THE COMPETITION. THAT HAS BEEN CONSIDERED THE WAY TO WIN. TODAY AND IN THE FUTURE, IN ORDER TO ADAPT AND EXCEL, THE INDUSTRY MUST UNDERSTAND THE INNOVATIVE MARKETING METHODS THAT THESE CHAPTERS ADDRESS.

You won't find palm trees if you decide to vacation in Pittsburgh, but if you are a pharmaceutical salesperson, you might indeed find the city much like Honolulu.

We learned this when working with a company whose branded drug had been doing poorly in Pittsburgh but really well in Harrisburg, where its market share was five times greater. The two Pennsylvania cities are only a few hundred miles apart, so the company had dispatched the Pittsburgh sales manager to Harrisburg to get some tips on how to set things right.

After meeting with the sales manager there, the Pittsburgh manager began implementing what he learned. Nothing happened. His business did not improve one bit.

A few months later, we did an analysis and learned that what drove the performance in Harrisburg was very different from what drove the performance in Pittsburgh. In fact, virtually everything the Harrisburg manager suggested was a bad idea for Pittsburgh.

It turned out that of all the cities in the nation, Honolulu was the most like Pittsburgh in terms of what drove the business. Why in the world would that be the case? You would think there would be an ocean of difference between them.

The analysis showed that Pittsburgh and Honolulu were both in a group driven by the status of managed care. Within that group, they most resembled each other in four aspects. Both had had a very similar, and poor, managed-care situation. In both, the physicians fell in the midrange, neither high nor low, in terms of writing prescriptions for the brand. Both had a low percentage of generic drug use. Neither was "high science," meaning they had fewer specialists for this specific disease state.

With such a profile in Pittsburgh, the greatest opportunity was in overcoming the poor managed-care situation. Such was also the case in Honolulu, despite the thousands of miles between them.

Pharmaceutical salespeople who serve physicians in the same geographic area tend to have the same boss. For example, all of Pennsylvania might have the same manager, who reports to someone who is in charge of the East Coast. None of the sales force in Pennsylvania would have contact with the salespeople in Hawaii and would not think to call them, even if the Hawaii folks could teach them the most.

Under traditional marketing, a pharmaceutical company works with an agency to polish its message, coming up with several points to differentiate its brand. It identifies who should get that message and then educates a sales force and sets it loose nationwide.

And as a result, the company gets a 30 percent share in Santa Barbara and 12 percent in Atlanta. Why would this be when the brand is the same and the salespeople all received the same training and use the same field tactics? Nationwide, we see major variations.

For the typical brand with which we worked from 2007 through 2010, the variation between lowest share market and highest share market was a multiple of four. From 2010 through 2014, that variation increased to a multiple of seven. The trend is clear. In just a few years, the discrepancy has widened dramatically because health-care is becoming more and more localized.

Jean-Luc Pilon, an industry executive with over 20 years of phar-maceutical marketing experience with companies such as Merck, GSK, Sanofi, and Vivus, put it like this: "Markets are not homog-enous; regionalization helps adjust marketing efforts to nonnational conditions."

The crux of regional marketing is to acknowledge and understand those variations so a company can do its best wherever it goes and wants to grow.

Healthcare is a local business built upon local trust, from peer to peer and from doctor to patient. It is based on what's happening right outside your door and your relationships here and now. What's happening in your backyard has a major influence, including policies that influence access and whether a prescription is filled as written by the local pharmacy and filled by insurance.

THE FAILINGS OF ONE-SIZE-FITS-ALL MARKETING

The one-size-fits-all approach has been the tradition since the dawn of the pharma industry. It has been done for simplicity's sake, but it no longer is the best approach because of all those variations from region to region.

Usually, the sales reps are the ones looking for those variables, with a bag of tools for different types of physicians. The reps size up each doctor in their territory, identify the early adopters, the simplicity seekers, and so on, and learn the prescribing habits and other characteristics of each doctor. Then they reach into their bags for the "best" tools.

The issue, according to Tim Cole, a regional sales director for Sanofi, with over 34 years' experience in the industry, is that all the reps are pulling from the same bag. Cole believes that reps in different parts of the country need to have different tools in their bags, depending on their local situation.

Those who develop the marketing message strive to differentiate the brands. They work hard through qualitative testing to find out why individual doctors choose a drug. The reasoning has been that you can roll out common-denominator messages across the country and that doctors will gravitate to the right message.

That doesn't work because it ignores a tremendous amount of information that is quantitative. It ignores the patient population. It ignores the prevalence of a disease in a region. It ignores the accessibility of healthcare and how government policy affects the writing and delivery of scripts.

The traditional game plan was to capture the heart and soul of the individual doctor by coming up with compelling differentiation. Regional marketing is a way of identifying the local drivers, understanding the many variations, and reacting to them effectively.

THE SIX Ps

The best sales reps understood their doctors and the environment in which they practiced. For example, they could say, "Hey, Dr. Smith, in your practice you have lots of patients who have Aetna as their healthcare insurance. Great news, my drug is covered by Aetna for only $5." The super-reps would know that not only was it covered by Aetna but also, they could add, "Dr. Smith, you also have lots of Hispanic patients."

In other words, they would be able to tailor their message. They would wrap it in the national messages, but they would include the local perspective of "this is what our drug does in your environment with those conditions."

To do this right, however, you need to look at more than just Aetna and Hispanics. Regional marketing looks at what we call the six Ps.

✦ Payer: Who's paying the bill? Is it the US government? Is it the employer? Is it paid in cash? Who's paying the bill can affect what goes on in a particular market.

✦ Provider: That means, essentially, the hospital systems. There are a variety of provider system metrics that can affect how a brand is marketed in a particular area.

✦ Prescriber: How many doctors are in a market, and what is their nature? Are there many specialists? Nurse practitioners? Who are the key opinion leaders? These all vary from market to market and can affect how receptive people are to the brand and message.

✦ Product: What are the special qualities of the brand, and how do they compare to the brands of competitors?

✦ Place: What are variables of the locality, such as policy issues and the frequency with which a sales rep meets with the doctor? In some states, smoking in public is illegal. If you are marketing a smoking cessation drug, that policy will affect your product's performance.

✦ Population: What are the local demographics? What percentage of the population is old, young, black, white, rich, poor, and so on?

In the old days, all that the great reps needed to communicate was the key message of the brand: that it was safe and effective, and whatever else they might throw in, such as that it was covered by a certain insurance company. But now it is essential to consider those six major variables, those six Ps, each of which also has 15 to 50 metrics. There's no way a sales rep can understand, for example, 150 metrics and figure out what makes a market different.

Marketers need a much more sophisticated analysis. Despite the complexity, through regional marketing, there is a process to analyze those 150 metrics and synthesize them to figure out what is most important in a particular market and to identify similar markets so that they can be grouped and managed effectively.

The task calls for an experienced regional marketing expert who can work with the brand team to figure out which approach makes the most sense in a particular market. A rep cannot just cut and paste the sales aid, trying to suit the doctor of the moment. The variations in markets are too many and too wide, and there's just way too much information for one sales professional to process.

You can be a fantastic sales rep and know your market well, but you don't have a perspective on other markets. You would not know, for example, that your region has six times as many specialists as the average market. You just assume that what you know is what everybody else knows.

In the not-so-distant past, that sales rep was also selling only one drug or a couple of drugs in a category. Now, a rep could be responsible for multiple drugs in, for example, a women's health portfolio, covering everything from overactive bladder to menopausal symptoms. Not only do you now have more doctors to touch, service, and understand but you also have to deal with different disease states and therapies and insurances. The burden worsens.

A TAILORED APPROACH

Here's an example to illustrate the significance of regional marketing and how the tailored approach works. In working with a drug for high cholesterol, we discovered five local drivers for how well it performed in each local market.

Some regions, such as the Atlanta area, have an abundance of patients with multiple risk factors. The patients not only have high cholesterol but hypertension, diabetes, and obesity, and they are smokers.

Other cities are high-science markets. They have more cardiologists, more teaching hospitals. Think of cities such as Philadelphia and Boston that have more doctors per population than elsewhere in the country.

We describe other markets as brand friendly. These are markets where far fewer generic brands are being used.

Then there are markets that we call "compare and win." To win in these markets, it is most important to establish the clinical difference between you and your key competitor.

Finally, there are regions that we call the Hispanic markets, where the focus has been on physicians who treat more Hispanic patients.

Each of those elements is important anywhere in the country, but if you are in Atlanta, you want to focus on the multiple-risk-factor patients and on getting that message to the physicians. In Philadelphia, you take the high-science approach. For any given region, you focus on what is known to drive sales there.

If we conclude, then, that Philadelphia and 12 other markets, such as Chicago, St. Louis, Seattle, and Santa Fe, fall into the high-science market, we can group them together into a virtual region. Then once we focus on what makes those markets unique and what drives a product's performance, we can educate physicians accordingly.

The conditions of the virtual region determine the best approach, in other words. For example, a certain disease is more prevalent in one virtual region. Specialized equipment is available in another. A centralized treatment center is located in another. One market has mostly primary care doctors; another has mostly specialists.

The traditional style of marketing presumes that it is impossible to tailor approaches to so many variables. The regional marketing approach affirms that it is absolutely possible if we look at markets throughout the entire country and group the similar ones.

This is not a geographical region approach, as in a "Northeast market" or a "Pacific Northwest market." Similar markets very often are not contiguous. Think of a map strewn with confetti, with a string of red connecting some markets and a string of yellow connecting other markets. Those similar markets need to be addressed with the same marketing tools.

There are many colors to the confetti, and in working with a company's brand team, we need to identify which drivers help and which hurt the product's performance, which are most actionable and why, which can be combined, which are most urgent, and which are most affordable to pursue. We identify the regional drivers of the product that are most important to the company.

If you are that overburdened sales rep trying to juggle many tools and messages in connection with a particular drug, you can now look at the regional market analysis and know how you will be more successful: in addition to understanding the individual doctors, you can focus on specific aspects that have been identified as drivers for the region.

IDENTIFYING SIMILAR MARKETS

How do we identify, then, where the confetti should fall on the map? When you consider those six Ps and the few hundred variables they entail, the task can seem daunting. The answer, in short, is to analyze a wealth of data so that the resources can best be allocated by virtual region.

Imagine an enormous Excel spreadsheet with 200 columns. The first column is brand share. Another column is prevalence of a disease. Others include the amount of insurance copay and the percentage of African American patients. There are a great many other variables.

Those are the columns. The spreadsheet has rows too, of course. Each row is a state (or city or sales territory)—Alabama, Alaska, Arizona—all the way through Wisconsin and Wyoming.

We look at that first column, the brand share, and our goal is to figure out whether there is a correlation with the metrics in the other columns. So we set to work, analyzing the statistics and applying the math.

Next, we look for how any of those columns correlate to total sales or brand volume. We check them out one at a time. Then we adjust the brand sales by population. Places such as New York and California don't automatically "win" just because they have more people.

To see what we might find, we do such things as multivariable regression analyses and random forest regression analyses, and we triangulate, market by market. In the end, we might identify ten drivers of performance, and we take that discovery back to the company's brand team to figure out which ones are most actionable.

multivariable regression analysis is

Typically, we narrow the number of drivers to a half dozen or fewer, which we compare, using a clustering methodology, to the drivers of all the states or metropolitan areas or sales districts or other types of factors we want to use to do the analysis.

You might think we would often unearth something stellar that nobody had imagined. In reality, what often comes out of it is more of a confirmation of all the messages, tactics, and strategies that the brand team has already defined. Many of them are right on.

"You don't want to cut turkey with a chainsaw," a long-ago mentor was fond of saying. He was pointing out the wasted effort of the traditional, uniform approach. Sometimes, that is called the peanut-butter approach: you spread it evenly, coast to coast. You need more finesse than that.

Now, we can say, for example, "There you are in Tuscaloosa, Alabama, and you have 17 tactics. How about focusing on these four? Of the three dozen slides in your presentation, you only have a few that are core to your business in Tuscaloosa."

That is liberating and empowering. When you get down to the most salient and compelling points, you will connect faster and deeper.

DOING MORE WITH LESS

We must do more with less, particularly in a world where regulation is now demanding it. In the era of reform, those involved in the healthcare industry, whether doctors or marketers, have been feeling the pressure to show that they are driving outcomes. The doctors and nurses and others in the field are expected to accomplish more with fewer resources.

Marketers too need to move the needle for greater effectiveness, and that is hard to do with a one-size-fits-all approach. To respond efficiently and effectively, they need to pay attention to those market-by-market variations. It should be good news to any brand team to hear that it can allocate what it has and that, by focusing its strategy, it can get better results.

When you get down to the local level, you uncover things that you have yet to address. You are going to have to deal with local policy, local access, local disease state prevalence, and patient population and do so in a scalable way. You need to produce materials common to your brand for all of the actionable districts, but you will have areas where you can add localized information. It is fairly quick to implement.

All of this comes down to an efficiency play. Companies often do nationwide TV commercials for a brand, but in some markets, the brand is not covered by the insurance company that has most of the patients. Let's say that's Idaho, and instead of costing $25 a month there, the product costs $100 a month. Still, having seen the ad, the Idahoans ask their doctors about the brand and find it costs four times as much as the competitive product. Few in Idaho are going to use that product. The resources spent on the advertising in Idaho were wasted.

We did an analysis on a cholesterol medicine and found that there were some markets in the country where adherence was one of the biggest issues for the brand. Most patients who were supposed to be on the medicine for the rest of their lives would stop taking it after. In some markets, adherence was even a greater issue: People stopped after only a month or two.

In the analysis, we overlaid those markets with ones that had a higher disease prevalence in general. That way, we saw markets with two traits: more people with high cholesterol and people with that same condition who quickly stopped taking the medicine for it.

This resulted in a radio commercial that went something like this: "Do you or someone you love have high cholesterol? I am Dr. John Jones, a cardiologist right here in Atlanta. I understand the importance of taking your cholesterol medicine. Failure to take it regularly can lead to heart attack and stroke." So it was a very local message that played only in markets with a larger adherence problem.

To make the message resonate even more, we used a local doctor to deliver it. If we found that St. Louis also had an adherence problem, we would get a St. Louis doctor to deliver the message. It is a targeted message based on the analysis. And the commercial doesn't play at all in areas with good adherence, so you don't waste money there.

It comes down to this: Why not use only the right messages delivered the right way by the right sales reps in the right districts at the right time? In that way, you should get the highest return on investment.

DOWN TO THE TACTICS

Once a company understands the importance of regional marketing and identifies the best markets for a product, how specifically does it take action?

This is where we get down to tactical development. After the analysis uncovers something not previously addressed, the company can produce brand materials that are specific and focus on the most important elements in a particular region. For example, perhaps the earlier brand materials made no mention of the Hispanic market, but the analysis shows that it is a major player.

The new materials will highlight that fact, while sharing a core foundation with other similar regions. Whatever has already been produced for the brand will be leveraged, including the messages and graphics and story lines. However, the materials are also going to include plug-and-play areas for localized information.

That is completely scalable. It is completely on brand. It utilizes a significant amount of content that has already been developed for the brand, so we know that it can be leveraged quickly.

Let's look at an example of how this all would work. After the analysis shows a higher percentage of Hispanics in, say, San Antonio, we put together a sales aid for the rep to use. It includes a lot of local data. At the top of the page is a drawing of the state of Texas, with an arrow pointing to San Antonio. Below are three bullet points: 1) 41 percent of Hispanics across the country have this disease state; 2) in San Antonio, it is 62 percent, which is 50 percent higher than the national average; and 3) that means that right here in San Antonio there are 612,484 patients who have this disease.

And so the sales rep goes in to see the doctor. "So what do you have new for me today?" the doctor asks. Traditionally, the rep might think, "To be honest, doctor, I don't have anything new for you today. I am trying to sell you an eight-year-old drug and convince you why it is better." But now, armed with this new sales aid, the representative says, "Doctor, I have some interesting information about what is going on right here in San Antonio," and explains the facts.

"Wow," the doctor says. "I didn't know this. I knew we had a lot of a Hispanics here, but I never realized our disease prevalence was 50 percent higher." The doctor feels engaged. He's getting relevant information, and so he's listening intently when the rep concludes, "That is why you need to use my product."

To expand this scenario, a lot of brand teams might say, "That is a great idea. Did you know in Brooklyn there are a lot of Russian patients? So let's do the same thing for the Russian patients." But there's no data to prove there's a correlation between the percentage of Russian population and brand share, as we established for Hispanics in San Antonio. So we should not be worrying about it. Around the country, there are 15 markets that are more than 15 or 20 percent Hispanic. There are probably only three that are more than 20 percent Russian. So that is not a very scalable market.

Cliff Barone, a brand marketing executive with over 25 years' experience on both the agency and manufacturer side, had a great insight on this topic: "Regional marketing provides the opportunity to better connect and drive greater impact in a smaller yet *scalable* way."

ALLOCATING RESOURCES WISELY

Every doctor in the country is a little different. The best way to market would be individual by individual, and you would need countless different messages, one for each doctor nationwide. It is too complex a task to consider in our highly regulated industry.

Even if we were to tell a brand team, "You have 15 different drivers for your brands, so we need to come up with 15 different marketing plans," the response would be, "Are you crazy? That is way too many." So we try to balance and learn, using the strongest mathematical correlation to determine which approach is most executable and scalable.

In our experience, up to half a dozen drivers are really all a pharmaceutical brand or sales team can effectively execute. So when our math points to 15 different drivers correlating to performance, we meet with the brand team and say, "Let's prioritize these. The math prioritizes it this way, but which ones really match your strategy?"

We were working on marketing for a bipolar drug for which psychiatrists wrote the vast majority of the scripts, and therefore, the sales team only called on psychiatrists. In the analysis, we learned that in about 18 percent of the markets around the country, a fifth of the scripts were written by primary care physicians. That told us whom to focus on in those regions: not just the psychiatrists but also the family doctors. The brand team acknowledged that but rejected the idea for practical reasons. It did not feel it had the salespeople for that approach. And so we went back and did further analysis and presented the second-best area of focus.

At a recent meeting, a marketing director pointed to a two-foot-high pile of resources on the edge of her desk that the brand team had produced for sales representatives. There was a four-page clinical reprint of study A and an eight-page clinical reprint of study B. There was the core visual aid. There must have been 40 tools.

Greg Pyszczymuka, a senior marketing director from Zogenix, with prior experience at Endo Pharmaceuticals, understands the issue of having too many resources. He believes a key element of success is making the most of existing resources and properly allocating them to best meet local customer needs.

In our analysis, we can take those 40 tools and say this, "In virtual region A, out of that two-foot-tall collection, you should use tools 1,

6, 8, and 12. In virtual region B, use 3, 4, 9, 16, and 25. You don't need to produce anything more."

Sales representatives have storage closets where they keep their product samples and the marketing materials. They have been known to have sizable trash cans there too. They quickly conclude that a lot of those materials, which took weeks (or months) to produce, just don't work well in their particular market. The materials in that pile of trash represent hundreds of thousands of dollars of effort. But the materials weren't appropriately allocated, so they ended up in the landfill.

Some of this is common sense. If you have a patient profile that recommends you do XYZ among your African American patients, you probably don't want to send many of those profiles to your sales representatives in Montana. You probably want to send a higher proportion to the sales reps working in Washington, DC.

That is a simple example that drives home this point: Allocate your resources wisely. What drives a brand's performance often is less obvious, something hard to see among dozens and dozens of variables. The sales reps won't usually know these local drivers. The interrelationships are just too complex.

On the contrary, the sales reps will often just choose to use the resource that they like best, and that might have nothing to do with what is going on in their markets: "I like this one best because the picture on it looks like my brother." However, the people in that region might not look much like the rep's brother. You need to allocate resources based on the biggest driver of performance in your market. If you are just spreading out the peanut butter evenly and sparingly, your approach is thin.

The only things pharmaceutical companies tend to allocate are samples. They'll send the right number of samples to the right person because samples are very expensive. And sales representatives' salaries are very expensive. If you pay the average sales professional $100,000, and you have a thousand of them, you are paying $100 million per year for your sales professionals. If we can make them 10 percent better by allocating the right regional sales tools, we are pulling a lever that will make a significant difference.

TURNING THE *TITANIC*

Regional marketing needs to be incorporated in the early planning stages. The gold standard is to start using these methods as best practice from the beginning so that you are getting the most for your money. Most pharmaceutical companies approach the challenge piecemeal. They try to make some changes in their marketing and their analytics.

It is like turning the *Titanic*. Everything about the marketing machine has been set up for the traditional approach. It runs from the C-suite and the Chief Marketing Officer to the vice president of marketing to the marketing teams and their brand managers to the people who do the analytics for each brand. All of it is set up to find the common denominator nationwide and to roll out the marketing in the hope that an individual doctor, somewhere, will select the product.

That marketing machine was successful for a long time. People in the industry were trained to believe that success came from inventing better products based on better science and marketing them far and wide. But today's world is very different. Your products aren't necessarily significantly better from a science perspective. To win, you have to figure out other nuances. You must not turn a blind eye to the local factors.

To get the most out of regional marketing and help a brand succeed, a lot of people have to be on board. We encourage brands to try a pilot. Let's say a company has identified a problem market, whether Chicago or Schenectady or both. We can do a localized analysis so the company can see how the process works. If we do our job right in the pilot, it will be easy for everyone to understand how it could be applied across the country for the entire brand.

Many senior managers in the pharma industry, however, grew up in an era when better science was the only way. To them, our new marketing ideas can seem foreign. Think of the packaged goods industry, which is, perhaps, 20 years ahead of pharma in this regard. In the 1960s and 1970s, a company would come out with a product that actually stopped cavities. It was better than the other tooth-pastes. Or a laundry detergent came out that actually cleaned clothes better. It was not long before all the toothpastes stopped cavities and all the detergents seemed to clean clothes equally well. How much cleaner could you get? The manufacturers had to start focusing on the nuances of marketing to gain the edge.

Today, pharma must market to regional nuances. This is not to say that marketing pharmaceuticals is as simple as marketing laundry detergents. Dirt in one city is not that much different from dirt in another city, but disease prevalence varies greatly from city A to city B. That means local and regional marketing makes even more sense for pharmaceutical brands.

TIME TO TAKE ACTION

In today's world of marketing, you have to be unique and you have to effective, but you also have to be efficient, and that can be a maddening combination. These are three competing questions: How can I be effective? How can I be compelling? How can I do more with a lower marketing budget? In the world in which marketers live, it seems you can pick any two of those but not all three. Regional marketing enables you to choose all three.

How do you start? Find a regional marketing expert. Explain what you know. Before your expert even starts an analysis, describe the regions where you have problems. You may have a good idea about the cause of the problem, but the experts can do the deep dive for further insights on why it is happening. Give them a place to start, a spot that has troubled you, where they can run a pilot.

If you think regional marketing makes sense and is compelling, don't use it as a Band-Aid. Don't use it just to take care of the gaps to fix the problem spots. It is far better to utilize regional marketing from the beginning during brand planning so that you only put together the materials for your brand that are appropriate to each individual driver. From the start, produce just the materials you need for just the right places. Otherwise, your efforts could end up in the landfill.

A lot of parity products are out there on the market, and to win, you have to find the tiny differences that will set you apart. Understanding that healthcare is a local market is an essential step, and each chapter of this book is about other differences you can make.

"Regional marketing success is like an orchestra," explains Kristin Vitanza, a senior marketing director from Endo, who started her career nearly two decades ago at Sanofi. "Each individual instrument has its own sheet of music, timing, and purpose. And when they all play together, you get the perfect song. Each instrument is essential to making beautiful music. Each component of regional marketing is one of the instruments: analytics, competitive intelligence, market insights, sales analytics, marketing strategy, messaging, and sales strategy."

CHAPTER 2

DIGITAL MARKETING

by R. J. Lewis

THE INTERNET HAS RADICALLY DISRUPTED MANY INDUSTRIES—
MUSIC, NEWSPAPERS, TRAVEL, TO NAME A FEW—AND THE
TIDAL WAVE OF DISRUPTION SURGES ONWARD. PHYSICIANS
AND PATIENTS SPEND COUNTLESS HOURS ONLINE. BOTH
HEALTHY AND SICK PEOPLE WILL WEAR DEVICES THAT REPORT
THEIR EVERY STEP AND HEARTBEAT. FOR ANY INDUSTRY,
THE COMPETITIVE ADVANTAGE IN THE DIGITAL AGE LIES IN
ADAPTING AND EVOLVING. THERE ARE VALUABLE LESSONS
HERE FOR PHARMACEUTICAL AND HEALTHCARE MARKETERS.

A highly personalized marketing campaign for a well-known antibiotic brand involved online banner messaging that targeted specific high-prescribing physicians. Since my company knew the doctors' exact identities—their full names, specialties, and addresses—we could both list-match target these individuals and personalize banner messages to include the physicians' last names as well as the status of the flu in their geographic area.

The campaign was powered with data from a third-party provider that tracked outbreaks of the flu nationwide, broken down by zip code. For each zip code, we knew whether the incidence of flu was worse than the year before.

We customized the ad for each individual physician to include his or her last name, the status and severity of influenza in his or her practice area, and a different color of the banner background, based on the severity of the flu at that time. For example, Dr. Scott Kolander in Ewing, New Jersey, would see a red-alert online banner: "Dr. Kolander, the Ewing area is in alert status for the flu." With one click, he would be shown graphical trending data and year-over-year comparisons for his geographic region. Seven out of every 100 doctors were clicking on the banner—an unprecedented 7 percent response rate. If the incidence of the flu was low in a doctor's area, the banner would be green. It still would be personalized with name and location, but the copy would indicate a "normal" flu status.

That campaign was educational and informative, and it provided a valuable service to physicians. It helped them better manage the demands on their practice and help set expectations with both staff and patients. When clicked, the landing page gave the doctor the ability to order sales-representative-delivered samples in order to meet growing patient demand.

It all sounds almost futuristic. But this was 1996. The Internet, as we know it now, did not exist. Browsers and search engines were relatively primitive and not widely used. This digital campaign was on the closed system of Physicians Online, and I was working at the time with Arnel Rillo, who was then the brand manager for Ceftin at the former Glaxo Wellcome company.

"The thought behind putting together the Ceftin flu alerts," Rillo says, "was to maximize our sales during the flu season. We wanted to provide an access point for local sales representatives to meet with physicians to provide them with real-time data on influenza and bacterial infections. This was the first attempt at target-specific personalized messaging as a service to customers, leveraging real-time data via the new digital medium of online services. We knew when, where, and how long the flu season would last and could allocate resources accordingly."

This program of two decades ago is relevant to the trends and capabilities we are seeing today on the Internet at large. The power of digital technology, the ability to leverage data in the name of servicing customers, and our ability to target and personalize messaging has existed for a long time, even though the pharmaceutical industry is still not making full use of these competitive advantages.

If this type of targeting and personalization was possible 20 years ago, imagine what tools you will have at your fingertips in the years to come. These capabilities and more are available today on a wider scale since we've moved beyond the closed environment of online services and deeper into the open web of the fully connected Internet. As the pace of innovation accelerates, the open interconnectivity of the web is making it increasingly more cost efficient and effective to deploy data-driven marketing.

Big opportunities await those who are willing to invest in these capabilities and take the necessary steps and political risks required to gain internal support. With so much data at your fingertips to serve your customers, why just shout your brand message to the masses?

Pharmaceutical companies have so much data available that customers would find useful. By focusing communication on the customer and following the platinum rule of "do unto others as they would like to have done unto them," you build more than a brand. You build brand loyalty. Imagine if the customer service divisions of pharmaceutical companies were as big as their sales and marketing machines. Should they be? In an era of outcomes-based medicine, we will see a shift in investment from a "tell them" sales force to a "serve them" customer service division.

Glaxo Wellcome, now GlaxoSmithKline, was an innovative early adopter and first mover, which has significant advantages. When was the last time you ran a service-driven, customer-centric marketing campaign with a 7 percent response rate? What's stopping you?

DISRUPTION

Both the healthcare market and the practice of marketing communications are in flux. Customers are rapidly shifting their media consumption habits. Once, they received their information from radio, TV, and newspapers, and they networked at dinner meetings. Today they are spending more time on the Web, using mobile devices, and engaging with online social networks, which all involve consuming more digital content. Highly fragmented media has led to multitasking and limited attention spans.

In spite of the regulatory environment, the politics, internal challenges, and all the other countless excuses, pharmaceutical and healthcare marketers must face the realities of a new digital age.

The Affordable Care Act is here to stay. The payment dynamic is changing, with payments now tied to outcomes. Clinical trial data alone does not support the negotiation process with payers as it once did. Payers want data that reflects the "real-world" use of products. To be effective going forward, all participants in the healthcare ecosystem must work together and leverage digital technologies that support outcomes in order to deliver patient results.

In every industry, over time, inefficiencies are eliminated. This is the essence of the Affordable Care Act. Government is trying to pay for outcomes and results instead of for procedures and transactions. Ironically, pharmaceutical marketers themselves have done this for years with their own vendors, funding only those programs that deliver meaningful return on investment.

In digital media, for example, there are various ways to pay for advertising. There's CPM-based advertising or cost per thousand impressions delivered. There's CPC advertising, cost per click. Then there's CPL, cost per lead. There's even CPA advertising, which I refer to as cost per "anything" because payment is tied to any "action" mutually agreed to and delivered, such as viewing a video or signing up for a newsletter.

Over time, the trend between pharmaceutical companies and their vendors has been to push the pricing model down that continuum and pay less for the unguaranteed result and more for the guaranteed result. But now, the pharmaceutical industry finds itself, uncomfortably, on the receiving end of the "prove-your-results" conversation and is being asked to prove real-world results despite having limited control over many aspects of what drives an outcome in a complicated system with many participants.

Payers are pressuring for results, the actual efficacy of a course of therapy as opposed to simply paying for the number of prescriptions written. So if patients don't improve or fail to take their medications, pharmaceuticals are being asked to share the risk.

Pharmaceuticals' typical clinical-trial-backed-marketing approach touts the efficacy of their drug. But managed care is pushing back on clinical trial data because, by definition, trials are highly orchestrated and conducted in a near-perfect environment where the patients are evaluated and qualified and the drug regimen enforced to a specification. Payers today want to see supporting real-world data addressing how these medicines perform under actual conditions in which some patients take medicine incorrectly or simply stop taking them.

When a company produces the real-world data, the payer will say, "We want to see the clinical trials." If one is positive, they want to see the other, and if both are good, they want to do their own testing instead. This battle over access, combined with being held responsible for conditions and actions of others that seemingly fall outside a brand's control, can feel like a lose-lose situation. But successful marketers will learn to expand their influence of control and align with other constituents on delivering improved outcomes.

CHALLENGING THE STATUS QUO IN MEDIA ENGAGEMENT

While publishers of medical print journals are often cutting their frequency of circulation down, by contrast, high-quality digital influencers now publish multiple times each day, including on weekends, or better yet, they pursue real-time user engagement that drives community and loyalty.

Some examples of the newer disruptive Internet services that attract physicians and healthcare professionals include WebMD/Medscape, MDLinx, Doximity, QuantiaMD, Diabetes in Control, Drugs.com, and Doctors Lounge. There are also audience aggregators who deliver niche audiences at scale, such as my company, eHealthcare Solutions.

In the digital world, when one venue breaks a story, others often repurpose it for their own niche audience. The same or similar information is available in hundreds of venues now. It is hard to be the single source for information anymore. As rapidly as privacy is changing, so are the acceptable standards of fair use under the copyright laws. The balance of enforcement for a publisher is challenging because publishers want links back to their content. So, by vigorously enforcing copyrights, they also isolate themselves, which does not necessarily work well in a networked world.

Competition is intense. People's readership habits are changing. We ingest and consume information differently today. Nowhere is this more magnified than in the rapid shift to mobile devices. Customers tend to catch up on their news when they are standing in line at a CVS, device in hand with two minutes to spare, or while sitting in the waiting room at their doctor's office.

People prefer bite-size information today. Content has to be portable for all channels, not just websites and mobile but also for social media, e-mail, and other frequently consumed channels. The need for headline shock and attention-grabbing, as well as the need to write copy in a search-engine-friendly way, in order to win the love and respect of the search engines, is changing our language, style, and skill set. Unfortunately, to be successful today, writers write first and foremost for the machines, then for shock value to attract a short attention span, and then finally for purpose. Media knows that if they win the love of the search engines, the people will follow. The sheer number of machines crawling the Web has unintended consequences as well, such as false ad delivery being generated by "nonhuman traffic" (NHT) and the subsequent questions that arise as to was my advertisement ever viewed by a real person? Was the ad impression "viewable": a challenge the industry is only beginning to address and reconcile.

MISSED OPPORTUNITIES

Digital changes are inexpensive, providing the opportunity for constant testing and iterating. Amazon and Facebook, for example, both run hundreds of versions of their services at any given moment, testing to find out what works best. They constantly look at the data to assess which page layouts, features, icons, fonts, and color schemes deliver the best results, and they optimize the use of all of them. These changes result in incremental daily improvements to user engagement. Compare that to the pharma industry, where a company launches a website for its brand and then, often, doesn't do anything with it for a year or more. Rapid iteration represents a currently untapped competitive advantage for both pharmaceutical brands and medical publishers.

Digital also represents an excellent proving ground for any larger off-line program. Once you understand how your content and information gets consumed digitally, you can leverage this knowledge to improve the typically larger investment that must occur in off-line materials.

Historically, a medical journal might only publish a dozen or so articles that were reviewed and selected by a single editor or a small group of editors out of, perhaps, 500 submissions. Publishers should flip that thinking around. Dozens of the rejected articles that were close to making the cut were likely very good and, perhaps, even more relevant and useful to some of their constituents than those selected for print. All of these could be included online. Why not learn from Reddit, the popular news aggregator site that allows users to vote content up and down in the rankings, and let the readers decide what is most important to them and their practice? Why not make new content available daily?

Traditional publishers and marketers both must shift their thinking from "digital-first" to "mobile first", with a smaller screen in mind. This means imagining every digital creation first through the lens and power of a mobile device, as this is how it will likely be consumed most often.

It is a difficult mindset to change for the reasons that Clayton Christensen outlines in *The Innovator's Dilemma*. The incumbents of a business are trying to protect a franchise that is semi-functional, even if it is declining. They know how to operate it. They are comfortable with it. Their existing customer base likes it. They have ingrained beliefs about that business, but to be effective, they ultimately have to disrupt themselves and find the customers of tomorrow before someone else does. In other words, they have to shake up what they are currently doing. They have to take a page from Apple and "think differently".

FIGURING OUT THE NEW DYNAMICS

The pharma industry itself is an example of conservative forces trapped in the Innovator's Dilemma, trying to protect a franchise and preserve the status quo. At a digital pharma conference I attended, an executive of a major pharmaceutical company stood during the closing session to say, essentially, this:

> My head is spinning because the speakers here have such amazing and incredible, creative ideas on leveraging technology to help us grow our brands. But let me share my point of view. I oversee several multiple-billion-dollar brands. My number-one mandate is to protect the brand. I cannot have our company end up on the front page of the New York Times because we did something risky or overstepped an FDA guideline. My number-one mandate is to protect the brand. Number two is to grow the brand, but number one is to protect.

That is the mindset. Companies are cautious. Supported by a limiting belief that there is a lack of clear governmental guidance on how they can operate, pharma companies have been engaging in a conservative wait-and-see approach, with some toe-dipping trial and error to determine what they can do digitally. The FDA has actually come out with several rounds of guidelines as well as many warning letters that serve as a proxy for guidance. While FDA guidance could certainly be clearer, the agency has already answered a lot of questions. It is the pharmaceutical companies that remain unwilling to interpret the guidance and advance their marketing efforts to attain a competitive advantage. Most industries shudder at more regulations, guidelines, and guidance; ours seemingly clamors for more guidance from the FDA. Worse, we use the absence of guidance as an excuse for inaction.

Pharmaceutical companies now realize they must invest fairly heavily in digital efforts. Their big questions, often in this order, are: What are we allowed to do digitally? What are the restrictions? What are the opportunities? We should focus only on this last question: What are the opportunities? Then approach that discussion with the premise that "no" is not a viable response internally and instead should be replaced with "how can we?" Where there is a will, there is a way.

The opportunities are many. But to embrace them and evolve, pharmaceutical companies need to figure out how to become more comfortable with the ability to change. A company's ability to adapt is the greatest competitive advantage it can possess in a rapidly changing market. The successful companies that are causing industry disruption, as well as the ones who are surviving it, all possess the ability to adapt quickly.

PERSONALIZING AND TARGETING

We are in a marketing era of targeting and personalization. Pharmaceutical companies now have the ability to do more than just buy online advertising on a website. Now, when they buy advertising, they can specify a target list of, say, 50,000 doctors they want to reach. Third-party data sources can be combined to scale to a larger audience by targeting relevant content, by geography, or through look-a-like modeling: finding similar people with similar characteristics.

The ad can be personalized, for example, putting the doctor's name on it. The personalizing can be done in obvious ways, such as notifying the doctor that the area is on alert for the flu. Or it can be less obvious, such as providing comparative data to those who are researching a competitor's brand.

The marketing message can respond to users' past actions. If they already have looked up a product's efficacy, for example, the ad can call their attention to something else, such as the product's safety profile.

Though personalizing takes many forms, in general it leverages data on individuals and their experiences. It gauges their frame of mind while viewing. For example, is the reader researching or just searching for a quick reference? Is the reader relaxed and open to receiving information?

The ability to identify specific individuals online is really powerful, and the technology to do so is becoming readily available. Digital marketing provides a clear measure of how effectively the marketer's message is being delivered. Every interaction—every click, hover, mouse movement, coupon download, and engagement—can be measured and logged to fine-tune the marketing campaign and improve upon personalization. Marketers can clearly understand their return on their digital marketing investment. The data cannot be ignored.

Digital is an increasingly important channel to reach the customer because digital goes where the traditional sales force cannot. The number of sales representatives across the industry is in decline, and the number of "no-see" policies in larger health systems is on the rise. A few years ago, representatives could not effectively see about a quarter of their target physicians. Today 49 percent of prescribers are not accessible to representatives, according to a recent ZS Associates AccessMonitor survey. When your representatives do get in the door, they generally have about 45 seconds of face time with the doctor. And much of that time is spent securing her signature for a sample drop. Pharmaceutical representatives have become the most expensive delivery personnel on the planet.

Let me give you an example that illustrates the power of using digital for personal targeting. One of our clients has a relatively small sales force. The company set up a symposium in which a key opinion leader was to speak. Their sales representatives had been assigned the task of recruiting physicians to fill the seats but were failing miserably. We received a panicked call a few weeks before the event and were told their representatives had only confirmed a handful of doctors for the program. How could we help?

In days, we set up and executed a highly targeted digital campaign for them. Although the representatives had had months to sign up physicians, and we had only had days, we were able to enroll over three times the number of physicians the representatives had been able to and make the meeting a success. That's the power of nonpersonal digital promotion.

A SERVICE ORIENTATION

As pharma moves from its focus on products toward an orientation of service, it has an opportunity to leverage these online technologies. Companies can look not so much for individuals but for the context of their thinking: What are they reading? What are they searching? What did they recently find?

By understanding their mindset, you can shift the messaging into a helpful service tool rather than a product billboard. The message, in essence, is: "Hey, we are here to help. Here is just what you might need right now."

As a company shifts from being a brand-centric organization to being a healthcare service-oriented organization, some powerful tools it might use include personalization of the message, understanding what the user is seeking, delivering the message in a contextually appropriate environment, and adding data-driven value, such as the previous example of providing the status of flu in their geographic area.

Used together, marketers can deliver value to customers by putting the right message in the right context at the right time and delivering it through the best media.

THE "AUCTION-BASED" MODEL

As technology has enhanced the ability to target specific people, digital marketing continues to move toward an "auction-based" model. It works like this:

Pharmaceutical brands gain intelligence about who prescribes a drug and how much. The company uses that list to target certain physicians and purchases exposure to only this select group of prescribers.

Some doctors are bound to be more value for one company than for another, depending on its brands. Pfizer, for example, may be willing to pay more to reach a specific physician than, say, Merck, but Merck may outbid Pfizer on a different physician who prescribes more Merck products. In effect, they will bid the amount they are willing to pay, typically in CPM, or cost-per-thousand exposure format, and the winning bid gets the ad exposure.

For years, pharma has been buying search terms and keywords in the manner of Google AdWords, and it is comfortable with doing so. That too is auction-driven: the pricing is variable based on the demand and the supply. I believe that this model is going to grow in pharma marketing as well, when it comes to reaching their high prescribing physicians.

We are very close to the day when a pharma company is going to upload a list of 125,000 prescribers it wants to target and announce, "We are willing to pay $400 CPM to reach Dr. Jones, $280 to reach Dr. Smith, and $175 CPM to reach Dr. Abernathy"—individual bids to reach individual prescribers.

In terms of doctors' prescription writing volume in a given category, not all doctors are valued the same to a marketer. That is a reality that pharma has embraced for a long time. Ultimately, somewhere out there is the highest-value doctor for your brand. Marketers know that prescriber well. They court her and make sure she is supported. After all, she's the clinician writing the most prescriptions for your drug. She is clearly of very high value. Different values are then ascribed to prescribers, such as no-see physicians. Marketers want to reach them, but they cannot effectively message them any other way.

The segmentation of value per physician is coming. Decile- and quintile-driven marketing, breaking the target list into groups of ten or five equal segments of value, are just convenient simplified proxies to help us categorize individuals into manageable buckets. But when you can target at the individual level and dynamically personalize messaging, there is no longer a need for proxies. Your ability to buy media targeting individual physicians in an auction-driven value-based manner is coming. Will you be ready for it? Or are you still analyzing your data in deciles? Is your infrastructure set up to personalize both the bidding process and the messaging to individual doctors?

Your high prescribers are often your competitors' high prescribers too. If they are on your target list, there's a very good chance they are on many others' lists as well. As was the case when Ceftin got 7 percent click-through rates on the flu alert program, and when mobile first arrived on the scene, there will be advantages for first movers.

MOBILE MARKETING

While mobile is here for most industries, mobile is still a relatively new frontier for the pharmaceutical industry. Again, the companies that act first will gain big advantages. This is particularly evident with price: When there's nobody bidding against you, you can gain exposure relatively inexpensively. You have the performance advantage because the medium is new. It is hard to quantify the value of the expertise and learning gained by being a first-mover. But first movers clearly develop a competitive edge in knowledge, skills and adaptability. One executive we interviewed said that by 2015 at least 25 percent of pharma media is going to be mobile specific. At least that's a start since it's the dominant place customers spend their time today.

In the early days of mobile, the click-through rates were remarkably high, attributable to the fact that it was a relatively new medium. Now, the rates are much lower. How much did you avail yourself of the first-mover advantage?

Mobile is not going to be an advertising play, per se, as much as it is going to be a utilitarian play. This is a great example for which pharma could capitalize on its service offerings. Marketers today, for example, focus on building a mobile app to manage persistence and adherence: making sure patients are timely with their medications. Perhaps the app is part of a prescription "program" or a service that comes with the pill prescription and enables the patient to track the doses, reach out to other patients for support, or send a report to the doctor. The app could send reminder alerts for taking the medicine, and if a patient failed to log a dose, another reminder would get sent by text. A separate alert would remind the patient of a refill and even offer to automatically process the refill. The app might also include a feedback loop, offering to e-mail the doctor with any information the patient's app recorded before the next office visit.

Does that sound like something you have worked on or are working on? Suppose we apply the platinum rule and look at it from the customer's perspective. Wouldn't this be a much more useful tool for the patient if it handled all the customer's medications (not just your brand) and maybe sent a reminder of their doctor's appointments as well? It could notify patients if any of their prescriptions or physicians were not covered by their insurance plan, even if that occasionally worked against the best interest of your brand.

What kind of data do you think you could gather from this app? Do you think, in the long term, you would be in a better position, despite some short-term switches away from your brand that the app helped facilitate (which would have happened anyway)? How much more utilization would a user-friendly tool generate? How much more data? The good news and bad news is that Apple sees health as such a major opportunity that it is now supporting it centrally with the launch of HealthKit. How long do you think it will be before something even more powerful than this exists?

Covering all brands from the patient's perspective, not just yours, now makes this a valuable service for the patient as well as for the doctor. That is the utilitarian service aspect of providing patient support.

At present, Apple is focused more on wearable devices that measure bodily activities automatically. They measure and track your footsteps or your heart rate or other vital statistics and are getting increasingly more sophisticated.

Patients, both healthy and sick, self-report by wearing these devices, and we continue to see big advancements in both the data they collect and the range of what they measure. The consumer-oriented devices that do fun little things employ technologies that medical devices use, but they are improving and will measure things such as glucose counts, cholesterol, and how and when drugs are administered and absorbed. Roll the clock forward. If many patients can walk around wearing devices instead of entering the hospital for monitoring and tests, how might this disrupt the hospital industry even further?

As we enter the "Internet of things" where most household devices and appliances are Internet-enabled, how does this create opportunity for your health program service offerings? Will a patient's refrigerator ultimately act as a gatekeeper for late-night snacking and report who has been into the fridge and how often? Will the home scale send a reminder to patients to weigh themselves in the morning if they forget? Whom should you be partnering with to best take advantage of the Internet of things?

A SEISMIC SHIFT

As the healthcare industry changes dramatically, marketers must figure out how to be more relevant to customers. They must reach multiple stakeholders. Whether a patient takes a medication depends on so many factors. For example, a patient must first choose to go to a doctor, then the doctor must make a proper diagnosis, which might be preceded by a lab test, which the patient must get. Upon diagnosis, the doctor must write the script, the patient must take it to the pharmacy, the insurance company must cover it, and the pharmacy must fill it without calling the doctor and recommending a change to a generic. And then the patient has to actually take the medicine properly. With so many stakeholders and steps involved in the process, no wonder it can be a challenge to influence an outcome.

If you are selling shoes online, you can be reasonably confident how the sale happened and even track attribution and influence: after seeing your display banner six times, someone clicked on your Google keyword and bought a pair of shoes. But for pharma to get a prescription, all the above steps and often more need to happen. Any one step in the process can derail prescription fulfillment and often does.

There is the possibility of leakage at every step where a patient can be lost on the path to a prescription. For example, a doctor writes a prescription, and the patient later decides she isn't really feeling that bad or cannot really afford to fill the prescription and doesn't fill it at a pharmacy.

Digital marketing is not a perfect science, even outside the pharmaceutical industry. A display ad for shoes might lead to a purchase even if the buyer doesn't click on it but later looked up the phone number and called in an order. So who gets credit for the results? Attribution can be tricky to measure, even when you are selling shoes. Yet attribution will become increasingly important under the Affordable Care Act. When you are paid based on results, everyone willingly steps in to take credit for those results. Attribution measurement is a skill set that pharmaceutical marketers still need to perfect. Their future depends on it.

Pharmaceutical companies have many "customers" to please. To be relevant, pharmaceutical companies must be more than pill makers. They must also become healthcare service providers. How can they add value, beyond supplying the medication? That is where the industry can present data, analyses, and technological advances that will help it fulfill that service role.

There are many players in today's healthcare system. Only those who can add measurable value will thrive. The pharmaceutical industry has the opportunity to add value by servicing its customers and potential customers in holistic ways that follow the platinum rule. If the manufacturers fail to deliver for customers, I can assure you, others will. Expect insurance companies, hospitals, employers, start-ups and technology companies to fulfill these customers needs instead. Others will reap the benefits of brand loyalty and data insights while further distancing pharmaceutical companies from their customers.

By default, focusing on the customer means taking more responsibility for outcomes. What good was the pill if it wasn't taken or wasn't taken correctly or in the right dosage or simply didn't work due to other factors such as poor diet? How do you get more deeply involved to keep the patient from falling off of the path to a beneficial prescription? As marketers, you must make sure the patient recognizes the symptoms and goes to the doctor. You want to make sure that right lab tests get completed and the conversation leads to a clear diagnosis, and you want to make sure that when the circumstances warrant it, the doctor chooses your product to prescribe. You want to make sure that the patient gets the prescription to the pharmacy;

that their insurance covers it and they can afford the copay; that the pharmacist doesn't switch to a generic; and that the patient takes the medicine properly, refills it in a timely manner, and, most importantly, the patient's outcome improves. Contributing value on the path to results for each of these various customers and influencers will keep your brand relevant. With increasing noise and fragmentation of attention spans, simply shouting about your brand does not get you heard, but servicing your customers is always a sure way to grow sales.

TURNING AROUND THE MISTRUST

One of pharma's challenges is that both patients and healthcare professionals, generally, distrust the industry. Our industry ranks down with tobacco companies and post-financial-crisis banks on the reputational index scores with consumers and healthcare professionals alike.

Much reputational harm has resulted from publicity concerning drugs that were pulled from the market. The industry has been accused of burying clinical trial data, misrepresenting product benefits, improper sales techniques, and other misconduct. The general lack of responsiveness, nonparticipation in social media, and PR impotence has not helped to build public trust in the industry.

We pay a price for blaming the regulatory environment for our own inability to communicate effectively. People feel that pharmaceutical brands are absent from the conversations online. We are absent usually because of the industry's conservative nature and our regulatory environment, which leads companies to falsely believe they need to maintain silence beyond the standard recommendation of, "Ask your doctor if [our brand] is right for you."

Customers notice when brands are absent from the social media conversation. When they tweet a question about a brand, they generally receive a preapproved standard response. Painfully, this does not feel like quality customer service. If someone were to report an adverse event to the company, the pharmaceutical company has a legal obligation to report this adverse event to the FDA. The result is that pharmaceutical companies choose to shy away from such conversations, but they miss valuable customer feedback and engagement in doing so. One of the major unintended consequences of this behavior is that it contributes to the distrust of the pharmaceutical industry and our brands. We can do better.

Companies such as Lilly and GlaxoSmithKline have taken the lead in helping to repair the industry's reputation. Yolanda Johnson-Moton, Lilly's director of external relations for its US medical division, is working to communicate more proactively through a YouTube channel and through educational initiatives with healthcare professionals and consumers.

"Working in a complex healthcare environment requires collaboration and partnership among all key stakeholders," she says. "Building trusted and valued relationships between the pharmaceutical industry and the healthcare community is imperative if we are to drive solutions to improve health outcomes for people."

The education of what the industry does must be externally focused and also internally communicated. For example, we worked with Johnson-Moton on an online video to, initially, educate healthcare professionals, which has also now become part of Lilly's corporate training. More external and internal communication like this, on rebuilding trust, needs to occur, and actions speak louder than words.

The importance of these efforts to re-establish trust cannot be understated. It is a foundational necessity as we move into an outcomes-driven environment. To change behavior requires a significant level of trust. After all, you are not the patients' employer, insurance company, spouse, or caregiver. Why should they trust a pharmaceutical marketer? Before marketers can expect that trust, it must be earned.

A few years back, refusing to accept "doing nothing" as an option, and instead asking "how can we?", Lilly challenged itself to find out what it actually could do in social media. The company decided to identify the major bloggers in relevant therapeutic categories and invite them to visit Lilly as guests. Lilly representatives spent some real quality time with these influencers.

A company spokesperson engaged the bloggers in dialogue and let them know that, indeed, the company reads their blogs and gets a lot out of value from them. The spokesperson explained that the company empathizes with the comments made about their drugs and that they pay attention when people ask the questions, "Where are the brands in these conversations? Why aren't they participating?" The spokesperson really got to know the bloggers and explained the regulatory environment in which the company operated and why the company could not be more active in responding to patients directly. They discussed the FDA regulations and the adverse-event reporting requirements, in addition to the federal Health Insurance Portability and Accountability Act (HIPAA) and privacy issues and the need for doctor/patient confidentiality. The spokesperson explained the host of reasons that Lilly could not be as active in social media as the bloggers and their readers might desire or expect or as the company itself wanted to be.

The bloggers responded enthusiastically to Lilly's willingness to listen to them and explain its vulnerabilities and challenges. They left with a much deeper understanding of the pharmaceutical industry's perspective. They thanked Lilly's spokesperson for taking the time to explain, discuss, educate, and listen. When they returned, these influential bloggers started posting positive feedback about their experience. They explained in their own words, with sympathetic language, the regulatory environment that pharmaceuticals were facing and why they often seemed to be missing from the conversations. This built trust.

What turned things around? Lilly found those bloggers via social media, but at the end of the day it was the personal touch, meeting face to face and listening closely that made the difference. There are things we can do, even in a highly regulated environment, to build trust and have a meaningful positive impact.

PRIVACY ISSUES

Privacy issues are bound to be debated as we move into a world of personalized, targeted messaging. To some, this world seems outright spooky. But that resistance is starting to disappear, particularly among young people who seem to value privacy in a different way and instinctively understand the tradeoff between privacy rights and the exchange of value for free and relevant information and services.

For the rest of us, these new marketing methods might feel like an intrusion on our privacy. But the market is evolving, and these sentiments are rapidly changing. As marketers, we need to evolve our own beliefs. In a rapidly changing environment, we must frequently revisit and re-evaluate our policies to ensure they are still relevant as we move into the new era of pharmaceutical marketing.

In general, the older you are, the harder this is to accept. Young people's expectations for privacy today are quite low. For the newest generation, it will be even lower. My two-year-old, for example, has been on video, often recorded for posterity, since the day he was born. We have a baby monitor at his crib. Many people go a step beyond this and now use devices to keep track of an infant's heart rate, breathing, temperature, and more. Some are embedding tracking chips in both pets and children. Many people have cameras throughout the house. When we leave, traffic cameras and surveillance cameras as well as individuals with smartphones can record our whereabouts whether we know it or not. We are nearing the time when any of this critical information can be routinely shared with parents, pediatricians, and caregivers in real time to monitor for abnormalities as well as general safety.

Cameras are proliferating. George Orwell's prognostications in his book *1984* were not too far off. Moments off camera will be rare for today's kids as they grow up. "If you see an unattended bag when traveling, alert the authorities." This is the new normal. What kind of expectations for privacy do you think they will have when they are in their twenties? As we adopt and embrace technological benefits, the perception of privacy is dramatically changing, and marketers must change with it.

I consider myself web savvy and contemporary, but I still get a chill up my spine at times when I see how my information is being used. Before I got married a few years ago, I had been shopping on Blue Nile, a jeweler website, for an engagement ring. At one point, my soon-to-be fiancée and I were sitting on the couch, and I looked over her shoulder as she used my laptop to look something up. I started to sweat bullets because I could see the Blue Nile ads popping up on every site she visited.

She looked up at me. "Wow, Blue Nile must be having a really big sale," she exclaimed. "Their ads are everywhere." Who said advertising doesn't get noticed? Thankfully, she did not put two and two together.

That was four years ago, when interest-based, or "behavioral", targeting was in its infancy. Now, it is a lot more prevalent, we are all a lot more web savvy, and it happens so often that you are bound to get the feeling that someone is following you. This targeting is becoming pervasive because it works for the marketer and the consumer. Despite that episode, I bought her diamond at Blue Nile.

Target, the store, does a lot of analysis of its shopping data. Target has been able to determine, based on the customer's age, gender, buying habits, and other factors, when a woman is likely pregnant. And Target uses this data. For example, there are certain things that mothers-to-be tend to start buying, such as prenatal vitamins but also less obvious but still telling purchases like skin creams, when they become pregnant. The data shows this fairly clearly, and data analyzed correctly doesn't lie.

One day, an angry father walked into a Target store and in essence asked, "Hey, what the heck is going on here? You sent my daughter this circular full of baby items like Pampers and all of these baby products. I don't want you to give her any ideas by sending her these kinds of things. She's still in high school."

The local manager, likely unaware of the corporation's marketing efforts at the time, apologized profusely. Two weeks later, the father returned and confessed to owing the store manager an apology. After a heart-to-heart with her, the father found out that indeed his daughter was pregnant. But what he really wanted to know was how Target could possibly know that before he did. Target's data-driven marketing engine had picked up on the patterned buying habits and had marketed accordingly.

Those kinds of stories illustrate that we still clearly need quality marketers, real people with common sense, to direct our technological efforts and ensure appropriate use of data. This story also illustrates a generational difference in attitudes. Once people grow up experiencing personalized marketing as normal, having been exposed since birth, they will increasingly see such communication as a service, not a threat. We are slowly but surely eliminating waste from our promotional spends. Eventually, consumers won't want, or expect, to see products that are irrelevant to them. After all, if you are a 75-year-old man and you get a Target circular featuring infant Pampers, unless you are a recent grandfather, are you really likely to purchase? And if you are a recent grandfather, Target will know that.

Interest-based advertising, in which people are targeted based on what they have viewed in the past, is a highly sensitive subject in consumer healthcare. The Network Advertising Initiative (NAI) created guidelines for what's appropriate in healthcare. (See http://www.networkadvertising.org/2013_Principles.pdf.) Unfortunately, these guidelines are open to subjective interpretation of what is considered "precise" targeting, which requires different levels of opt-in. The guidelines state, for example, "All types of cancer, mental-health-related conditions, and sexually transmitted diseases are 'precise' and require opt-in consent. Other conditions such as acne, high blood pressure, heartburn, cold and flu, and cholesterol management, the NAI considers to be generic and not topics that require opt-in consent." Everything in between requires subjective interpretation of the guidelines.

The NAI created some definitions and guidelines. Appropriateness has to do with the prevalence of the disease. It has to do with common knowledge and acceptability and whether it is something that an average person would consider to be particularly private in nature, which again is highly subjective and varies from person to person.

Much of this is personal choice. If I were suffering from diabetes, I might have a great deal to say about the disease. I might have no qualms in telling everybody about my condition because I am a vocal diabetes patient advocate and educator. I may want everyone to be aware of the symptoms, so I speak openly and frequently about my disease. However, other diabetics might view their condition as a private and confidential matter and not want it disclosed. The utopia may, indeed, be each consumer's personal choice. Healthcare privacy will likely evolve in this direction of user choice, as other services have. As Craig De Large says,

> "There was never going to be online banking, right? Because no one was going to put their most sacred commodity, money, at that kind of risk. Remember that? Now we don't think twice about online banking. Over time, the media and the financial services industry have re-educated us. In a similar fashion, we're going to also see this happen with health privacy. I believe that user-controlled privacy settings I now have with a service such

as Facebook will ultimately take hold in healthcare in a similar way, at an individual, user-controlled, level."

We get questions all the time about data ownership from both publishers and marketers. These are tricky subjects for both marketers and publishers to navigate effectively, especially on the patient side, but taking the end-user view is often very helpful; how would you want your information handled?

On the professional side, with full disclosure to end users, publishers help eHealthcare Solutions to authenticate physicians when they register for services online. That information, when matched against known databases of prescribers such as the Drug Enforcement Administration database and the National Prescriber Identification database of the Centers for Medicare and Medicaid Services, allows us to "authenticate" a known prescriber. Once we authenticate prescribers, we are able to target them when a marketer asks to reach them specifically. Doctors have been targeted individually by brands for decades. After all, prescribers are the customers of the brands whose prescriptions they write as well as of the publishers whose content they consume.

Despite a "customer" status, privacy is still a very personal matter. We debate with advertisers and publishers alike over "who owns the customer data." I believe we are moving toward a world in which each individual owns his or her own customer profile. Marketers, publishers, database houses, and service providers of all sorts simply have the honor, privilege, and responsibility of housing that data and using it responsibly, politely, and in a manner consistent with the customer's wishes. This is, indeed, an awesome responsibility. We have seen what has happened recently with several high-profile customer data breaches, such as the Home Depot credit card breaches. These events are devastating when it comes to customer trust.

The notions of data ownership have to change. Individuals, ultimately, own and should control their own data. The rest of us are just stewards of that data, seeking the opportunity to leverage the information to deliver a better customer service and better customer experience.

Data-driven digital marketing is a service that society will grow comfortable with over time, just as we have become comfortable with online shopping and online banking. By targeting only those who are likely to want the service, you avoid wasted marketing dollars and you stop wasting your customers' time.

In other words, efficiency saves time and money for all, which, in turn, drives the ability to better service customers. Quality marketing messages of the future will be less about advertising and more about delivering timely, appropriate, and useful resources, services, and notifications.

Facebook is a great example of a company that lives on the cutting edge of privacy. It is constantly getting its hands slapped by a government agency or privacy watchdog group, and every now and then, even its users revolt. So Facebook pulls back just a little bit, the service improves, and people again become happy with it. Facebook continues to grow, and then it pushes the privacy envelope again.

If you have never read the Facebook or Google terms of service, read them. If you can understand the vague language, you will be surprised at the potential rights to your privacy that you are giving up by using these services. For example, Facebook retains a royalty-free right to use and reproduce everything you ever submit. This sounds logical, given the nature of how they need to display this information to others when you use their service, but think about the potential ramifications of that broad right. Google records and collects your activities, such as search queries, links you click, and YouTube videos you watched, and the company may combine this data to assemble a user profile on you. The trade-off in both of these instances is that you receive something highly relevant and useful, such as finding useful information or connecting with people you have not seen in years. But free services come with a trade-off. When you are getting a digital product for free, it is important to remember that you (and more precisely your data) *are* the product.

The more you reveal about yourself online, whether it is to Facebook, Google, or whomever, the more relevant, customized, and valuable you become to marketers and their paying customers and the more valuable their offerings become to you.

As a result, society is rapidly evolving its views of acceptable data practices, and privacy, as we know it, is diminishing.

At the time of this writing, 23andMe, a company that will map your genome for a small fee, announced it was going into drug development. Consumers can opt out of having their data used in research. How much will those suffering from a terrible life-threatening disease for which there is, currently, no cure value privacy over the ability to make a difference by contributing their genetic data to the research of a cure? Privacy is a new form of currency that is willingly exchanged for a product, service, or even the hope of a cure. As service offerings become more valuable, our willingness to trade our privacy in exchange for them increases. However, as with any supply and demand curve, our privacy becomes less valuable as the supply of those willing to trade off their privacy increases.

Information has always been a form of currency. Linked with big data analytics and improved targeting methods, information is rapidly becoming the most lucrative form of currency. Information and our access to it and ability to use it is easier than ever before, and it is driving massive technical innovation, which is creating increased productivity and improved quality of life. Most importantly, it represents a monumental opportunity for pharmaceutical and healthcare marketers to better service and communicate with their customers. So what are you waiting for?

CHAPTER 3

PROMISE OF BIG DATA

by Brad Sitler

THE ERA OF BIG DATA, IN WHICH WE HAVE UNPRECEDENTED ACCESS TO AN AVALANCHE OF INFORMATION, CAN CHANGE FOREVER HOW WELL WE, AS A SOCIETY, MANAGE OUR HEALTH. BY TAPPING INTO IT, THE LIFE SCIENCES INDUSTRY HAS THE ABILITY TO HELP PEOPLE AND TO HELP ITS BOTTOM LINE IF IT CAN SUMMON THE WILL TO DO SO.

Every 15 minutes, we generate the same amount of data globally as we did from the beginning of recorded history through 2005. Not since Gutenberg's printing press has society seen such a momentous shift in our ability to record data. The printing press changed the world, but that hardly compares to the staggering potential of the era of big data we're in now.

A popular term these days, *big data*, in essence, refers to the enormous volume of information and our ability to access it and analyze it in ways never before possible, thanks to advances in high-performance computing. No longer are we limited to sampling small subsets to draw insights and piece together the big picture. In the world of big data, we come closer to actually seeing the entire picture as we navigate the avalanche of data.

Today we can store vast amounts of data at a much lower cost, and the information comes from a multitude of sources. Think about the volume of social media chatter and the increasing number of devices, such as activity trackers, smart watches, video cameras, and smartphones that capture, feed, and share data. The Internet of Things (IoT), with its estimated 25 billion connected devices that are rapidly increasing in number, will continue to drive this tremendous growth in data volume that is available for capture and analyses.

Some of that data is structured but most is unstructured. By structured, I'm referring to information that's shown in a chart or table; it's formatted and recognizable, arrayed in columns with letters and numbers. By contrast, unstructured data is what we find in places such as social media, in posts and tweets, and wherever else people leave information—whether an idle thought, a complaint, a shrewd observation, or just an emoticon. Unstructured data is free-form and untamed and yet, often holds valuable insights about attitudes, perceptions, and behaviors.

We know how to approach structured data; we have a history of analyzing it. But the analysis of unstructured data is relatively new. One would think that a smiley face attached to a message would be of significance only to the immediate recipient. Perhaps not. There are ways to tap into Facebook and Twitter feed data and the like, gather it into an analytics environment, and then mine it for valuable information. Today we have software that can glean the meanings from natural language. It can tell from the context, for example, whether the word "light" refers to luminescence or weight. The software is able to discern differences based on the structure of a sentence.

Further, the information itself is constantly changing, which makes it that much harder to understand and analyze. The percentages of structured and unstructured information in data feeds change regularly, with new types of data, not previously analyzed, introduced throughout the process. This makes the task of analysis more difficult but certainly not impossible. In fact, the pharmaceutical industry is already harvesting more refined insights from this mixed and massive volume of data.

Fortunately, new data management capabilities can handle these enormous and complex data sets. There are analytics capabilities for the business analyst and visualization solutions for data exploration. More professionals than ever before will be able to make use of data because all corporate teams will work to identify business opportunities.

Empowering more employees with the right tools will allow for more and faster insights: "Analysts working on big data need a seamless integration of data management, data visualization, and advanced modeling capabilities that, in turn, enable them to identify the appropriate data, prepare the data, iterate on multiple hypotheses, and attain insights faster. Time to insights is decreased when these capabilities are brought together in an intuitive workflow and design interface for the user," says Jamie Powers, senior industry consultant and big-data authority at SAS.

SEIZING THE OPPORTUNITIES

Some people find it unsettling, or even frightening, that such a huge amount of information can be tapped to reveal so much about each of us. It may seem Orwellian, as if Big Brother is watching.

The industry needs to take note of that attitude, since the public is not particularly trusting of the industry in any case. Some people unfairly blame it for contributing to the rising cost of healthcare, while politicians malign the industry. It is clear that pharma needs to work harder to develop trust with many consumers and patients.

By accessing big data, pharma can gain insights on the market and how to best interact with consumers. It can find ways to battle those negative perceptions and turn them around. By analyzing the information that big data makes available, the life sciences industry has an unprecedented chance to change those perceptions in the marketplace.

Slowly, pharma has been embracing social media analytics and leveraging big data. With a herd mentality, participants are wading into the murky waters. The opportunity is significant for gaining insights on branded products, therapeutic categories, and competitors' brands. Pharma can encourage feedback, both positive and negative, from patients and providers. In doing so, it can address issues that arise and protect corporate reputations.

Unfortunately, unfounded concerns linger around the reporting of adverse medical events that surface during social media analyses. Actually, less than 0.01 percent of adverse events on social media meet the requirements for reporting. Nonetheless, much of the industry has yet to participate in this social listening and to gain real advantages of basing decisions on hard data rather than on speculation.

In addition, other more impactful opportunities are available to pharma with big data. For example, companies can assess patients' ability to manage their own health and can enable and support them in doing so. The data can help the industry understand where resources should be targeted for the best outcome and greatest good for a patient population.

Pursuing these opportunities requires pharma to be more assertive. Two things need to occur. First, the life science companies must, to a degree, reinvent themselves. This involves redefining their commercial model of what they are selling, whom they are selling it to, and how they go to market.

Second, to reinvent themselves they need to foster changes in the regulatory environment that allow for collaboration with other entities across the healthcare continuum. Current regulations and the ultra-conservative way in which the companies interpret them have been roadblocks to substantive changes in the commercial model that would allow pharma to capitalize on big data.

IDENTIFYING PATIENTS AT RISK

As a senior healthcare consultant, I work in an innovation center, created by SAS, a large software company based in Cary, North Carolina. At SAS we looked at what was happening in the healthcare market in this era of reform and created an incubation and prototyping team, the Center for Health, Analytics and Insights (CHAI). Earlier in my career, I was an independent consultant to the pharma industry for a number of years. Before that, I worked for a large database marketing company, managing three of its largest biopharmaceutical clients.

At SAS, I deal with the three verticals in healthcare: life sciences companies, providers, and health plans. CHAI started with 17 people recruited from those three verticals, with the goal of developing software capabilities with which the industry could prioritize allocation of resources.

Our team focuses on developing ways to identify which patients are at highest risk of a medical event. To a large extent, this also involves an analysis of patient adherence (how likely patients are to continue to take their prescription therapy as their doctor prescribes).

For example, let's say two patients, Tom and George, have the same chronic disease, are in the same stage of disease progression, and are on the same treatment regimen with the same doctor. However, Tom is predicted to be twice as likely as George is to have a critical care event or progression of the disease. The doctor and health insurer will want to allocate more resources—more hand holding, if you will— for Tom, knowing that George's risk profile is lower and that George appears to be more self-sufficient. George doesn't need a nurse calling to urge him to show up for his appointment next week. He doesn't need a care manager or his health coach to remind him to walk three miles every other day. But Tom needs those calls.

At SAS, we're focusing on developing software to help life sciences companies, healthcare providers, and health plans to identify such at-risk patients, and to develop psychosocial profiles for those patients so that resources can be best allocated in a manner that is meaningful and effective. The goal is to develop the engine that selects the right intervention for each patient at exactly the right time to enable a sustained behavioral change. There is a breadth of new data assets that are now being analyzed to better understand the many facets of patients and their behavior.

BIG DATA SOURCES

Let's look at some specific sources of big data and how we can make the most of the available information. In short, here's what is needed: the data, of course, which will provide the insights; an IT infrastructure; analytical software; human resources (meaning the people who will gain the insights); and a clinical or business operation to use those insights effectively. These all must be put into place, and that can readily be done now. The data is valuable in understanding, predicting, and ideally, changing patient behavior. The smart money in this era of US healthcare reform is being placed on changing patient health outcomes.

With the volume of personal information available today, some wonder whether the data can be mined for personal tidbits or just for trends. Of the types of data that I list below, some can be licensed at the individual level. Other information is only licensable provided no specific individual is identified, which means someone can see your data but cannot identify you, personally, as the origin of such data.

And what are the data sources for the healthcare industry? Here is a rather comprehensive list:

- ✦ Socioeconomic behavioral data, obtainable from companies such as Epsilon, Experian, and FICO.
- ✦ Prescription claims data, which is information about the scripts you fill at your pharmacy. Companies such as IMS Health and Symphony Health Solutions provide that data.
- ✦ Electronic medical record (EMR) data is all the information that physicians, pharmacies, or hospitals collect about a patient that can shared in an EMR.

✦ Lab data, such as information from diagnostic tests.

✦ Patient registries data: hospitals or providers can capture information about patients with a particular disease and track how their conditions progress over time.

✦ Social media data coming from sources such as Facebook and Twitter.

✦ Health plan claims data is data from a health insurer that has a good view of prescriptions filled and diagnostic tests performed. This information does not show results, which are captured in the lab data and the EMR.

✦ Medical device data, such as from a blood glucose monitor: the sensor data indicates, for example, how frequently patients check their blood glucose level and the results of those tests. Additionally, there are a dearth of new biometric devices streaming data (e.g. blood pressure monitors and wireless scales).

✦ Health device data: today there is a multitude of mobile technologies such as activity trackers, smartwatches, and fitness apps for smartphones that monitor and capture data. Fitbit and smartwatches will produce significant amounts of activity data into the future.

✦ Patient support or intervention program data is collected from the interaction a patient has with the program, such as communications with a health coach or text message reminders to fill a prescription.

By combining these data sets to create a composite profile of a patient both inside and outside a clinical setting, we can identify untapped opportunities to administer appropriate care. We will start to see the impact of environmental, psychosocial, and behavioral aspects to disease management and be able to proactively stratify patients based upon risk of a future medical event.

The final source of data on the list, patient support or intervention program data, is becoming increasingly important and will drive specific interventions for specific patients. A major healthcare focus will be on behavioral change. How do we get patients to change their behavior, such as follow a better diet, exercise more, or take their medication twice a day after meals? Historically, the healthcare industry has not been very successful in driving patient behavioral change, as evidenced by the adherence rates in treating obesity and type 2 diabetes, both of which involve lifestyle factors.

Medical personnel have been telling patients to lose weight, exercise, and eat better, but the industry has not done a good job of assessing what works and what doesn't in changing behavior and for whom it works. For example, if we can determine that peer support groups work best for people with a certain profile, we should work harder to establish such groups for those people.

To help crack the code on patient behavior change, we need to combine and mine the data sources to drive targeted interventions for specific patients. This a sizable part of the US$40B population health management market projected for 2018 (Markets and Markets, Population Health Management Market, 2014, report). The "patient behavior change" market has been referred to as the next billion-dollar blockbuster (drug) market. Patient behavior change is a core component of the population health management market. If we are unsuccessful at supporting sustained patient behavior change, the goal of healthcare reform will not be realized. Life sciences is interested in the population health management market for two reasons:

1) changing patient behavior will include getting patients to be more adherent, which is a tremendous problem for life sciences, and 2) life sciences are trying to figure out how to have a meaningful role in population health management since it will significantly change our healthcare system.

THE MAJOR CHALLENGES

The biggest challenge for the life sciences and pharma industry is in driving innovation in this area. Using big data, the industry can identify opportunities to deliver something new to the market where there is a need.

With healthcare reform, life sciences companies will need to show an improvement in patient outcomes and show that what they are offering is cost effective. They'll need to prove it is a better outcome for a patient at a particular cost versus something the competition is doing. Health economics and outcomes research (HEOR) groups within life sciences companies do this today. They are working with big data to identify insights that quantify the outcomes and cost-benefit value of their drug versus that of the competition.

Market competition has increased on two fronts. First, many of the blockbuster drugs have gone off-patent. From 2010 to 2015, an estimated $130 billion in branded drug sales was lost to generics. Generics are forcing life sciences companies to step up their research and development. But many of the new drugs are for smaller patient populations and hence are demanding significantly higher prices. One example is the 12-week regimen for hepatitis C at a cost of over $80K. Although this particular drug does have a strong financial-to-outcomes proposition, the sustainability of these types of pricing models is in debate. Considering that the rest of healthcare is moving to fee-for-value, one would suspect they are not sustainable, especially when other treatment options are available. This will push the industry to reinvent its business model, although at this time, most of the companies have not.

Another challenge for pharma then, is to build that new commercial model. I have talked with a number of innovation teams at pharma companies. These folks devise good, insightful ideas. But historically, the organizational culture has not embraced and invested much in those new ideas because they have not had to do so. Pharma has been incredibly profitable for half a century or more. Once a company secured a molecule under patent and started selling in a less competitive or exclusive market, it could often take a price increase if it was concerned about making its revenue number the next year. Why then, would a company turn its business model upside down when it was always going to satisfy Wall Street? The industry had no incentive to change its business model. However, in other industries, big data has enabled a few entrants to redesign the entire market— think travel, with Expedia and Priceline; think financial, with eTrade; and retail, with Amazon.

A third major challenge is to change the thinking of mid level pharma management to harness change. For example one pharma company explored new commercial business models but was unsure whether its employees, some of whom had been in the industry for 20 years, would adjust and get on board with new roles and a new direction. Would the staff have the entrepreneurial mojo to forge ahead both internally and externally?

It doesn't appear that the company ever made any substantive changes. It's unclear whether its legal advisors slowed down the initiatives, or stopped them, or whether there has been concern at the very top of the organization about fundamentally changing the business model or expanding on innovative plans.

WAKE-UP CALL

Nonetheless, the fact that companies are setting up innovation centers shows they recognize the potential. The industry needs to be aware of how reform measures are altering the nature of healthcare. We are moving from fee-for-service—the system whereby physicians get paid whenever they see a patient or do a procedure—into a fee-for-value system, under which the providers earn money by ensuring that the patient gets better. If the patient doesn't get better, the provider and/or payer makes less money or doesn't get paid.

This is a major industry shift. As health plans and providers take on more financial risk through their contracting, they will seek to share that risk with other entities across the healthcare value chain. It's very likely that pharma will soon find itself facing the same value scrutiny in the US that it is already facing globally.

With the creation of health information exchanges and other comprehensive healthcare data repositories, the holistic view of patients' treatment pathways in a real-world setting will empower providers and health plans with insights into drug and therapy performance. In crowded therapeutic categories, the formulary gatekeepers may further reduce coverage on high-cost therapies when lower-cost alternatives with equal clinical outcomes exist. These concerns are emerging across the industry.

"Imagine aggregating say cholesterol levels of every patient who's taking your statin drug and really understanding how the drug is working in the real world. We would need to control for all of the factors that need to be controlled for on a large scale like disease severity, age, comorbidities, and these types of things. But this big data set would allow us to really understand if the drug is really effective, and if not, the data may perhaps explain why. "Real world data is scary to some companies because it may not show that their product is best. In reality there's no such thing as best, there is only what is best for an individual patient.

—Lars Merk, Portfolio Marketing Director of Diabetes Franchise, AstraZeneca

And with advances in software and computing power, there are new high-performance analytic solutions that support the analysis of tens of millions of rows of data in a fraction of the time that it formerly required. Additionally, advances in data visualization allow for exploration to realize more insights even faster. The big data sets combined with these advanced analytic capabilities start to set the foundation for precision medicine. Precision medicine refers to the tailoring of medical treatment to the individual characteristics of each patient. This is accomplished by segmenting a patient with a particular disease/condition into subpopulation cohorts that differ in their response to a specific treatment. Patients within cohorts receive the treatment regimen that has been optimized for their particular circumstances.

Given the scenario presented above for the status drug, what if negotiations with providers or health plans turned into small, branded, volume purchases for select, targeted cohorts of patients who are predicted to not respond to other lower-cost therapies?

The rapid development of the breadth and depth of data to enable analyses like these has been noted by a leader in big data insights: "We leverage big data capabilities to create actionable intelligence," says Tom Henry, vice president of knowledge solutions for Express Scripts. In serving the pharmacy needs of nearly 90 million members, he says, the company generates demographic, environmental, psychosocial, behavioral, and health data. "Our advanced analytic capabilities produce unique insights that help patients and clients make better decisions and achieve healthier outcomes."

Unfortunately for pharma, challenges are also emerging from entrants outside the industry. Because the changes within US healthcare are so significant and truly daunting, other industries that were not previously involved in healthcare are seeking a role in helping patients or providers make that behavior change and improve outcomes.

The lesson is this: If pharma and life sciences companies don't step up, they may be left out. At this time, at least four companies serve as examples.

✦ Apple recently launched HealthKit, its new software platform for collecting data from various health and fitness apps. This composite view can include data such as heart rate, blood pressure, body temperature, weight, blood glucose, distance walked, steps climbed, and other data from the myriad of apps from third-party developers. HealthKit also has been designed to share data with provider organizations. Imagine the volume of data streaming from Apple's 73 million iPhone users in the USA as they upgrade to the iPhone 6 with HealthKit. It will be a treasure trove for providers or any company looking to enter the patient engagement or population health management market. Who would have thought that Apple would participate in healthcare? But if Apple does successfully enter the field and become a competitor with anything your company is doing or could do, you should be concerned. Why? Because Apple understands the value of leveraging big data and disrupting markets.

✦ Sony is entering the medical device and diagnostics market. In short order, diagnostic imaging data, streaming medical device data, and a portfolio of smartwatches will position Sony as another major entrant in healthcare with access to the big data trove.

✦ Ford has been working on capturing consumers' biometric data. Sync, an on-board health monitor, captures a driver's biometrics and makes recommendations on driving routes to accommodate the driver's need to stop for an insulin injection. Additionally, it can send phone calls to voicemail so the driver can focus on driving in times of high stress.

✦ Google, in one of its top five strategies, is helping consumers and patients better manage their health by launching a free trial of live chat with doctors. This is done when a consumer conducts an online search for symptoms. The data from search history, if combined with the text from the video chats, can yield tremendous insights into predicting the spread of disease, helping triage patients faster and less expensively, and getting high-risk patients to a provider more quickly.

With such giants now entering the healthcare arena, the pharma industry should take notice. The competition is intense, immense and moving quickly. Apple and Google are companies that understand the concept of big data. They have developed new industries with disruptive technologies. They have reinvented industries and are dominating them, in large part due to their understanding of the potential of big data. Amazon certainly gets it, as it predicts what you might buy based on your profile and what you have bought in the past, as well as what people with similar profiles have purchased. That concept could be applied incredibly well to driving changes in patient behavior.

And these companies have figured out how to scale things. The iTunes store and Amazon are now looking at delivering fresh groceries. Both Apple and Amazon are incredibly adept at scaling their organizations into new markets. Life sciences companies today don't very often work directly with patients. To do so, they would need to find a way to scale their operations very quickly. There's much to learn from the consumer giants who are adept at catering to consumers very well.

This is a wake-up call, and pharma needs to hear it.

It is not that pharma companies don't want to do more; it's that they have not been motivated because they have been incredibly profitable. And again, the industry is naturally conservative, partly because of regulations and how the company's legal teams interpret them.

I recently attended a big data conference where a biotech representative offered some insight into the company's challenges. Essentially, the company wanted to provide more targeted support to patients and wanted both the patient and doctor to get involved actively.

A company representative visited Capitol Hill to seek a change in regulations and get some FDA feedback.

I'd like to see the industry do more of that. If a pharma company comes up with innovative ideas that appear to run counter to current regulations but that benefit the patient, provider, and others in healthcare—while keeping costs down, I think it should send people to Washington, D.C. to make their intentions clear and advocate for change.

MAPPING THE PATIENT'S JOURNEY

The life sciences industry has spent a fortune on research to understand how patients conduct their lives. It has done a tremendous amount of primary and secondary market research to develop complex patient segmentation models within most of the major disease categories. These segmentation models are valuable assets that can work in tandem with big data.

Segmentation models can be based upon defining patient attitudes, needs, and behaviors and how all three of those relate to prescription drug adherence. The models are being developed so life sciences companies can identify which people can be persuaded to change their behavior and adhere to therapy. The companies want to know where their investment of patient support resources will do the most good.

Other market researchers have looked at understanding a patient's journey from the initial symptoms to the decision to see a doctor to going for tests to returning to the doctor for a diagnosis and coming to terms with that diagnosis and starting a treatment plan. The patient comes to a resolution or seeks a second or third type of treatment or ongoing treatment if the condition is chronic. The research maps the clinical, mental, and emotional aspects of the patient's journey.

One reason to map that journey is to identify opportunities to engage patients when they are particularly receptive to more information or behavior change. The idea is to reach them at a critical juncture with information that might drive them to ask their doctor about a particular type of treatment or option.

These are some of the richest insights into how patients live with specific diseases and the daily challenges they face in balancing demands of family, work, and community. That knowledge provides a context for managing yet another daily demand of an acute or chronic condition. Pharma understands where, when, and why patients fail to adhere to their treatment regimens and make sustained lifestyle changes.

EXPANDED OPPORTUNITIES FOR PHARMA

Let's take a closer look at the opportunities for the pharma industry to take advantage of these unique assets in this age of big data, which can serve many roles in helping to reduce costs and support patient behavior change.

About two-thirds of our healthcare bill stems from chronic diseases, many of which are related to lifestyle. Life sciences companies could benefit greatly by helping providers to get their patients to change their behaviors. That could be as simple as getting them to adhere to therapy, or it could involve helping them change their lifestyles through exercise, diet, or better sleep.

Data analysis allows us to identify patients at greatest risk. Looking at a combination of EMR data, both structured and unstructured, combined with rich psychosocial data, my team is risk-stratifying patient populations so that resources can go to those patients who would benefit most. Specifically, we are seeking to identify patients on a trajectory for an admission, a readmission, or the progression of a preventable disease. If we can identify the at-risk patients, health plans and providers can intervene early enough to be effective. They can make deliberate levels of investment in support, based upon a patient's risk for medical events and associated costs.

However, our work goes much further. It is not only to identify patients at risk; it is also to create rich profiles and clusters of patients to support the design and selection of interventions that will work. If you are going to try to change patients' behavior, you must interact with them in a way that is meaningful to them. After all, if an intervention either does not reach a patient or the patient does not engage with it, the investment is for naught.

To do this, you need to use all the data you can access about that patient and similar patients. You can develop a robust understanding of the individual patient and the others in that group and thereby predict what might help to change that particular patient's behavior. The provider or life sciences company or health plan can then execute various tactics of support and intervention and can do so when the patient is most receptive and the intervention can do the most good.

Then you assess whether the behavior changed. If it did, you would start using that same tactic, or tactics, on other patients with similar medical profiles, psychosocial profiles, and behavioral profiles. Meanwhile, you would continually gather data during the interventions and support measures. Think of it as continually executing, assessing who was helped the most and who was not, and refining. How are those patients different from each other, and how are they similar?

Consider how your pharmaceutical company can get into the business of providing support programs to patients directly and take advantage of the frontline provider-patient relationship. Taking it a step further, consider how you can monetize these support programs. How do you charge health plans for those programs, where historically, you might not have wanted to charge, especially in more competitive therapeutic categories.

One of the major challenges is getting more data about the patient; you'll want to know whether a certain provider has a particular patient who is female, 33 years old, with a particular socioeconomic behavioral profile. There are ways to develop rich data repositories that can be used to set up support and intervention programs, without the life sciences company ever knowing that this is Jane Doe at 234 Locust Lane.

These challenges are significant, though not insurmountable. There are ways of doing this, but they have not been accomplished yet, nor has the full potential been realized.

OPPORTUNITY: HELP PATIENTS HELP THEMSELVES

To improve healthcare, one of the things that must take place is increasing engagement between patients and their providers. Let's say a nurse practitioner sees a diabetic patient on a quarterly basis. The practitioner tells the patient to use a blood glucose monitor four times a day, and if the blood glucose is not within a defined range, the patient needs to self-inject with insulin.

Today, a patient may or may not do that. When I worked on some diabetes projects in the past and talked with a number of patients with type 2 diabetes, I heard most of them remark, "It is really hard to manage my diabetes exactly as my provider has told me to do it, but I do the very best that I can." Patients have a variety of devices for managing and checking blood glucose, but those devices do not "talk" to one another. Many of those devices don't give the patients easy access to their data or to their readings, or a glucose monitor might not store data so it can easily be viewed later. The devices don't export to an Excel spreadsheet easy, even if they have that capability.

So patients can have great difficulty in collecting their data, and it is nearly impossible for that data to ever make it back to the nurse practitioner who is asking patients to measure their blood glucose four times a day and inject a specific amount of insulin.

What if a pharmaceutical company or a medical device company were to make those devices Bluetooth-connectible so the data would be sent to the cloud? The patient could review the data with the provider during the next office visit. The data would reveal the blood glucose level, how many times a day the patient checked it, and how often the patient injected the insulin. If the patient has an insulin pump, the data would show whether and when the insulin was injected. Maybe the database could be integrated with patients' iPhones so that they could just snap a picture of a meal before eating it. Some app might convert pictures into a longitudinal graphic of caloric intake.

Imagine if a pharmaceutical company also hired a third party to host a data cloud that could sync with various devices or apps and collect and archive this data. This "cloud" could also include real-time analytics that would continually mine the data both at the individual patient level, specific cohorts, and across large disease populations. The insights from the analytics could then be shared with the patients. For example, a patient could be notified at 2 p.m. on Tuesday of his or her predicted failure to hit a target number of steps that day, based upon information collected about that patient's activities on previous Tuesdays after 2 p.m. This provides patients with an additional opportunity to plan the rest of their day to avoid getting to the end of the day and realizing they didn't make their goal.

This forecasting and predicting capability is largely missing in today's health apps, fitness trackers, and other devices. Information tracked by such devices and apps could also be shared between patients with similar profiles. A patient would be able to give the healthcare provider access to the data residing in this cloud. This sounds similar to what Apple is planning with its HealthKit app. From our discussions with diabetes patients and providers, both parties indicate that having access to longitudinal data could go a long way toward improving glycemic control. Patients and their providers could review the data together to identify particular days or times of the week to strive for improvement.

These are just "what-if" examples. But what if a company were to enter the business of helping a patient collect data and share it with the provider, so they could then work together to better manage the patient's diabetes? That could be a breakthrough for pharma.

Granted, this is a grand vision that would transform how healthcare is delivered, and devising a successful business model may be an enormous undertaking. There are, however, a number of possible options for driving revenue. For example, would patients pay the pharmaceutical company a subscription to capture the data and make it accessible to the providers and themselves even though the pharma company might never look at the information because of patient information rights? When a provider requests a patient service, would a health plan provide some level of reimbursement even if the health plan could not see the data?

A legal team may consider this heresy, but it's an idea that presents a significant opportunity for pharma companies. If Apple's strategy for the iPhone is to capture your temperature and heart rate and other information off the skin, I wouldn't be surprised if it developed a sensor on the back of that phone to measure your perspiration for blood chemistry. With Apple already working toward an EPIC EMR integration, these possibilities are not so far-fetched. Additionally, a high percentage of doctors have iPhones and iPads. An app that Apple is running could show a doctor a patient's data over a secure connection. This would serve as Apple's entry into delivering content to providers in offices not equipped with an EPIC EMR.

OPPORTUNITY: A RICH, 360-DEGREE VIEW OF THE PATIENT

As the models for healthcare change, the pharma industry can create another indispensable role for itself. After all, pharma companies have some of the richest patient insight data: the patient journey, patient segmentation frameworks, insights from executing patient support programs and the needs, attitudes, and behaviors of patients between doctor visits.

Providers get a 15-minute consultation with their patients. They see them during office visits as they treat them and collect data captured in the EMR. They are able to use that data to assess, diagnose, and prescribe a treatment plan for the patient. That is not an easy undertaking in such a short time.

That view tends to end when the patient leaves the office and returns to being, say, a 42-year-old mother who has three kids at home and her father-in-law now living with the family because his wife has passed away and he's no longer able to care for himself. She works full-time as an accountant. She's 27 pounds overweight, and she's struggling with type 2 diabetes. With her husband also working full-time, they shuttle their kids around in the evenings and at weekends, and unfortunately, they eat a lot of fast food and don't have much time for exercise.

The physician has just said, "You're overweight. You have diabetes. You need to exercise more. We are going to put you on Metformin at first, and I want you to exercise three times a week, and I want you to lose 15 pounds. Thank you, and call the office if anything changes."

Given her busy life, you can imagine how she might have a difficult time suddenly incorporating four hours of exercise into her weekly schedule and cutting out fast food. It's basically a setup for failure.

But pharma companies have interviewed those diabetic patients and understand the challenges they face outside the doctor's office. They understand these patients have tried to lose weight when their doctor told them to do so. They understand these patients are frustrated, and as soon as they get their diagnosis, they already feel they've given up. They don't see any light at the end of the tunnel. How are they to find a way to include exercise in their weekly regimen, change their diet, and get their family to help them?

Pharma companies have seen all of this. They understand it, and they know that just telling a patient at the doctor's office to "go do this, this, and this" rarely works. I think that, to a certain degree, providers also know that it probably won't work. They only have 15 minutes to consult, and they are doing the best they can to identify, assess, diagnose, and define a treatment pathway.

Now, suppose the pharma company partnered with providers to: 1) get patient data into the cloud so that only providers could see it at the individual level; 2) enable the joining of EMR data with rich psychosocial data and remotely collected data; and 3) include patient behavioral-based segmentation frameworks that indicated attitudes, needs, and behaviors of patients relative to managing their health.

These very rich patient profiles would enable the providers to understand the patient's challenges outside the office setting and select specific interventions for specific times during the patient journey, based upon the exact timing of events, such as initial diagnosis, treatment initiation, first script, using the actual EMR data as timing triggers. The timing for these interventions could be further optimized for patient segments based upon prior patients' behavior in response to the interventions. This capability already exists and is being utilized in other industries.

GSK, as an example, has partnered with providers to assist in creating a more holistic view of the patient within the provider's setting through technology. Phil Golz, Global Head, Commercial Innovation, GSK, has said,

Over the past five years there has been a considerable leap forward in the volume and quality of health-care-related data generated both by health-care systems and by patients enabled by technology advances in smartphones, machine-to-machine connectivity and high-speed networks, as well as the emergence of social health platforms. GSK is working with partners in multiple geographies and across different disease areas on initiatives to better understand how analysis of this data can be used to provide higher quality of care for patients. For example, in an effort to test how to best improve health-care quality generally and patient outcomes in the US, GSK, and Community Care of North Carolina (CCNC) collaborated to develop a predictive analytics tool, Care Triage™, which uses small data like prescription refill history and hospital admission/discharge data to identify patients at risk for hospitalization due to medication management problems. The purpose of this initiative was not marketing but rather, to help GSK learn about new value-driven care models, while also helping CCNC pilot new population health tools. GSK and CCNC learned that the predictive power of the tool could be achieved from small data sets, which could address existing interoperability challenges.

This could be a step toward creating a more comprehensive view of patients' data outside the clinical setting by incorporating data captured from activity trackers, apps, and social media, as examples.

OPPORTUNITY: ENGAGING PATIENTS USING A 360-DEGREE VIEW

This "what-if" scenario could be extended even further if a pharma company were to engage in disease management instead of solely drug adherence efforts. Providers today don't necessarily have the time, nor are they compensated, for ongoing patient management and support outside the office beyond a documented care plan and giving patients the opportunity to talk with their provider.

Pharma companies could be brilliant at this. They know a tremendous amount about patients' challenges. They've created support and intervention programs with marginal success, largely based upon limited data. Historically, one problem with these programs is that they work through open invitation. They conduct direct-to-consumer advertising to motivate patients to raise their hands and identify themselves. The catch here is that the ones who reach out to pharma companies to sign up for programs are the ones who are already very engaged with their health. They probably need the least help. It is the patient with the incredibly hectic schedule, the one who works full-time and has three kids and the father-in-law in the house, who needs the most help. This patient would benefit most if the pharma company and the provider were to align and say, "We are going to figure out a way to provide some support for you to drive substantive change." But those business models don't exist today.

This is how I foresee pharma companies using patient information they already have in hand from several sources: the patient journey, patient segmentation frameworks, and their experience in executing patient support and patient intervention programs.

This opens the door for innovation in partnering with provider organizations so that every patient who is diagnosed with, say, type 2 diabetes, is then enrolled into a patient support and intervention program that is specifically tailored to each individual's needs.

Unfortunately, patient support and intervention programs today are "one-size-fits-all." If a provider does offer a type 2 diabetes support program, it includes everyone with that condition, whether the patient is 65 or 22 years old or retired with grandchildren or a traveling consultant. Each patient gets the exact same patient support program, untailored to that patient's specific health situation.

This is a recipe for failure because how you talk to a 65-year-old who is retired and maybe sees the grandkids three times a week needs to be drastically different from how you talk to a 22-year-old who travels four days a week and works long hours.

Pharma doesn't do one-size-fits-all programs. Pharma, because of the segmentation models that I previously described, understands the concept of developing communication programs and support programs tailored to individual patients. The industry understands that some patients are at higher risk than others of not adhering. Pharma, however, has been less successful at intervening with different levels of resources due to regulatory and legal concerns over treating one patient differently from another.

When taking into account large patient populations, the scale and scope of data involved for these opportunities could be defined as big data. This is especially true if multiple providers' patients are combined into a single cloud repository. Imagine if providers could see their own patients and anonymous data about other providers' patients. It would be even more valuable if providers shared the data associated with all of the patient support and intervention programs in the cloud. This would also help the entire healthcare system to learn faster and more efficiently which interventions work for which patients.

This suggests that pharma needs to step up and find new ways to work with provider organizations. It could mean approaching the FDA to make a ruling or taking the FDA to task to say, in effect, "If you're not going to step up and make a ruling, you are to blame for preventing us from trying to innovate to help patients and drive down the cost of healthcare."

If pharma companies don't collaborate to provide patient support services around their brands, somebody else will, and fast. A public example of pharma entering this arena is Pfizer's strategic alliance with CliniWorks to develop a population health management platform including patient care capabilities. It will be interesting to see if this is solely a technology play to gain a better understanding of provider and patient interaction or if Pfizer is actively looking to extend a service wrapper around its brands. There have been a refreshing number of partnerships between technology companies, such as MC10, Qualcomm, and Google X, and pharma companies to address both disease management opportunities and remote patient monitoring. The industry needs to increase the technology partnerships to further drive innovation in capturing patient data, analyzing the data, and translating the resulting insights into actions that support patient's behavior change.

I think it will be providers who will take the lead in finding partners to build support for patients. If pharma doesn't solicit providers and health plans to develop partnerships to better support patients and promote education, providers will look elsewhere for that help. Pharma companies need to take a seat at the head of the table.

An increasing number of vendors are working to fill this role today. Google has added it to its strategic plan, and Apple has it in its Operating System 8. I think both of these companies already have the mindset that they are going to have a role in healthcare, and I see that role as one of helping patients help themselves. Part and parcel with that is helping patients capture data about how they are managing their health and their diseases.

If pharma companies wanted to change their commercial models, and if they could, now is the time. If they don't enter the market soon, it may be too late. A company such as Samsung could compete against Apple and have a good shot at penetrating the market. However, I can't see a pharma company competing with Apple. I see innovation and a new commercial model at the core of Apple. Could a pharma company compete with an established Apple in the healthcare market? Probably not. Could it get in there before Apple gets there and figures it out? I think so. Pharma companies already know more about patients. They have more information capital on healthcare overall. I think they have a shot and can make it successful.

PRIVACY IS RELATIVE

In my opinion, privacy won't be a significant issue in this age of big data so long as we continue to use the information responsibly to help patients. People are concerned, and rightfully so, that somebody may steal their identity and access their bank accounts or hack through corporate firewalls to get their credit card numbers. But in general, to date, there hasn't been widespread concern about socioeconomic behavioral data being used by healthcare providers to identify what treatments would work best or which patient support and intervention programs should be recommended.

It appears that younger generations have less concern about their data privacy on the Internet and what they choose to make available. Older generations lean toward being more concerned and prudent about privacy and data. But if data is used to successfully help patients have better outcomes, perhaps the public won't mind that the data was obtained from Facebook or Twitter. If you can deliver a significant value, and it's recognized you used the data responsibly, people likely will be comfortable with it.

One of the biggest risks for healthcare's reputation in using data stems from public perception of patient privacy breaches of EMR data. One of my biggest concerns is that these cases will arouse public suspicion to cause a knee-jerk reaction demanding that EMR data not be shared with anyone outside the specific provider or payer entity that collected it. Such a public reaction could deprive us of a great opportunity for good, through richer insights into patients. Partners and potential partners of providers would no longer be able to share data with providers. Providers and those who support them could be irresponsible, triggering such a reaction. Therefore, it is essential that all healthcare entities that manage or access data today must secure and use it responsibly and protect it from release.

Personal health information constitutes an immense amount of the most private information about a person. It must be treated with dignity and in compliance with privacy laws. As we look to put health data to commercial use, we must remember that the primary mission is to help others. When that value is demonstrated, the population may be less likely to push back.

WHERE THERE'S A WILL, THERE'S A WAY

How do we inaugurate a new era of pharma marketing in a time when US baby boomers are aging and their health problems are compounding?

It really comes back to embracing innovation and being prepared to change the historic commercial model to explore new ways to support patient behavior change. That may come in the form of supporting patients directly, enabling patients and providers to have a better exchange of information and data, and of course, helping providers help their patients change their behavior.

As the bulk of the US population ages and healthcare issues and costs abound, we need to take action. If as a society, we are spending money on healthcare, by definition we are not spending money on education. We are not spending money on social programs. We are not spending money on infrastructure. We are not spending money on our military. We are not spending money on new natural resource development or our energy sources. By spending so much money on healthcare, both directly and indirectly, we are limiting our ability to spend elsewhere.

Nevertheless, for a time we will need to overinvest in healthcare to move to a more preventive care model so we can avoid some of the high-cost, chronic-disease situations we find ourselves in today. Obesity has been shown to lead to type 2 diabetes. If we can stop the lion's share of patients' weight gain by supporting weight loss through exercise and diet early on, we may avoid many critical care events and the ongoing chronic management and costs of type 2 diabetes.

In other words, if we were to spend a little more on the front end, we may save much more on the back end. I think that is a huge opportunity for the US, as a nation, and why healthcare reform is structured as it is. Preventive care is being covered, and it most certainly should be covered. It's penny-wise to spend less money on patients before their problems worsen and cost exponentially more to individuals, to the healthcare ecosystem, and to society.

The era of big data can have a significant and enduring impact on our ability to care for ourselves, as a society. The pharma industry, if it keeps up with the times, has the capacity to play a central role in designing the future of healthcare. The question is whether it has the will.

MOVING TO THE FRONT OF THE PACK

Recently, I was at an excellent conference hosted by the Johns Hopkins Center for Population Health Information Technology. It was attended by tremendous thought leaders, executives from large provider organizations, payer organizations, government think tanks, and government representatives.

One of the presenters referred to the book *The American Healthcare Paradox: Why Spending More Is Getting Us Less* by Elizabeth H. Bradley, Lauren A. Taylor, and Harvey V. Fineberg. The authors mention the familiar chart that shows life expectancies around the world and how much each nation spends per capita on healthcare.

This has always bothered me. The US spends so much more than other countries, yet our life expectancy is only about middle of the pack. In fact, one developing nation has a life expectancy closely aligned with the US, yet it spends nearly nothing on healthcare.

By rerunning the analysis, the authors basically exposed what that chart does not show and what it should show. They said we need to look at social service spending plus healthcare spending and then assess life expectancy.

A lot of other countries spend a lot more money on social services, essentially making sure that things such as "food deserts" don't exist, making sure that communities have much more green space, and including ways in which people can exercise. These lifestyle attributes are built into the society as cultural fiber, making it easier for people to achieve more exercise, have better diets, and utilize social services that improve quality of life.

When you add social services spending to the healthcare spending and compare other nations to the US, the picture becomes clear: we are right in the middle of the pack on how much we spend per capita for both combined. Factoring in social services elevates the rate of life expectancy, or the ROI. In other words, it pays to prevent.

This emphasizes the importance of spending more prevention dollars up front to save healthcare costs down the road. You can help prevent obesity, thus lessening the type 2 diabetes epidemic, and subsequently, avoid the tremendous costs of strokes, amputations, blindness, and the many other complications of the disease. We could avoid all that if we were to spend money up front to help people lose weight and exercise more and have a better diet.

As we redesign healthcare, and in part, accomplish this by supporting patients' sustained behavior change, we will inevitably invest in patient support and interventions that are not effective for every patient. Overinvestment will exist, for a time, as we learn what drives behavior change at the individual patient level and become more efficient in allocating future resources to support the behavior change of other patients. This is the massive and critical opportunity for the healthcare and pharmaceutical industry, in which big data has a central role.

It is big data that is allowing us to gain better insights into what really works. The data is rich enough, and as we collect more of it, the dividends can be immense.

CHAPTER 4

EVOLVING SYSTEMS OF CARE

by Roger Zan, Joanne McHugh,
and Stephen Morales

AS OUR HEALTHCARE BECOMES MORE VALUE-BASED, A VARIETY
OF SYSTEMS OF CARE HAS EMERGED. TO REMAIN COMPETITIVE,
MANUFACTURERS NEED TO ACTIVELY SUPPORT THESE INITIA-
TIVES. THEIR CHALLENGE: TO TRANSFORM HOW THEY DELIVER
AND PROMOTE THEIR BRANDS. THEY NEED TO SELL MORE
THAN PRODUCTS. THEY NEED TO SELL WELLNESS—AND AT A
GOOD VALUE.

C learly, the Affordable Care Act has advanced significant change in our healthcare system. However, well before these reform measures, massive changes were already taking place across the country, and those changes will continue to evolve how we deliver healthcare.

As a largely employer-based model, the US health system is seeing incentives shift and new stakeholders emerge as key decision makers. Patients are being empowered to have more of a say in their healthcare, and thus, employers are shifting accountability—and cost—to consumers. Providers, in order to compete, are forming national or super-regional networks to establish the scale required to take on more risk, such as Vanderbilt Clinical Integration Network or Stratus Health Alliance. Decisions on care protocols in these new systems of care are driven by teams of administrators and healthcare professionals, leaving the individual physician with less autonomy. Commercial payers are not only acquiring physician groups but they are also moving to performance-based models, shifting the financial risk to providers.

These changes have been happening partly because of the pressure of expense: The US healthcare system is the most costly in the world, accounting for 17 percent of the gross domestic product, and is expected to grow to approximately 20 percent by the end of this decade.

As populations age with increased life expectancy, chronic health conditions, such as diabetes, hypertension, cholesterol, and HIV, are also increasing. This has put significant strain on healthcare systems worldwide, although the United States has been weathering it poorly in comparison with other countries.

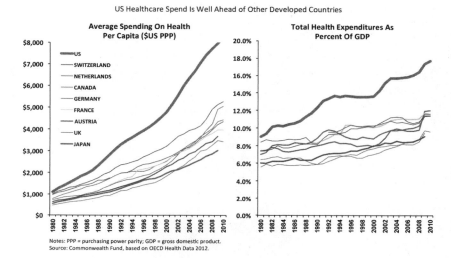

In effect, the US healthcare system doesn't sufficiently address the amount and type of demand from the US population. The amount of demand will continue to rise due to the aforementioned reasons, and the healthcare system also needs to more effectively address, clinically and economically, the chronic nature of diseases while improving our ability to manage the costly acute episodes of care.

THE TRIPLE AIM

The multifaceted objectives for health systems comprise the *Triple Aim,* a term coined by Donald Berwick, former administrator of the Centers for Medicare and Medicaid Services (CMS) and former president of the Institute for Healthcare Improvement, and adopted by CMS and other healthcare stakeholders. The Triple Aim centers on:

- Improving the patient experience of care (including quality and satisfaction)
- Improving the health of populations
- Reducing the per capita cost of healthcare

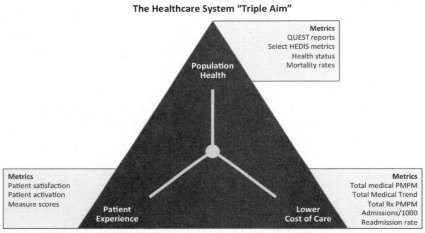

The Healthcare System "Triple Aim"

Metrics
QUEST reports
Select HEDIS metrics
Health status
Mortality rates

Population
Health

Metrics
Patient satisfaction
Patient activation
Measure scores

Patient
Experience

Lower
Cost of Care

Metrics
Total medical PMPM
Total Medical Trend
Total Rx PMPM
Admissions/1000
Readmission rate

Sources: (1) HRA: Understanding IDNs Report; (2) IHS: OneKey Market Insight Report; (3) HSG: Institutional Profiles; (4) Stanford: Closing the Quality Gap; (5) Navigant: ACO Strategy, Design, and Implementation; (6) Navigant Internal Interviews; and (7) Navigant Analysis

In order to deliver on the Triple Aim, systems of care are reconfiguring at an unprecedented rate, consolidating and integrating both horizontally and vertically. Horizontally, they are acquiring more of the same: hospital groups buying up other, often thinly capitalized hospital groups. This is creating strong national and regional systems with centralized functions. But they are also integrating vertically: they are buying tertiary-care centers, long-term-care facilities, and home-healthcare companies. This is purely indicative of the desire and need for economies of scale, in which larger organizations are better positioned to respond to the challenges of the current landscape than smaller ones.

These new configurations range from loose affiliations to tight affiliations. The models that are growing most quickly are Clinically Integrated Networks (CINs) and Accountable Care Organizations (ACOs). CINs allow various independent providers (hospitals, physicians) to collectively contract with payers and employers on the basis of enhanced quality and cost outcomes. ACOs are a group of providers that collectively contract to deliver services to discrete populations with accountability for predefined performance in quality and total cost of care. CINs and ACOs can be related, but they are not the same. An ACO is one way in which a CIN could contract with a payer, while a CIN is one method of physician alignment that could be leveraged by an ACO.

CLINICALLY INTEGRATED NETWORKS (CINS)

CINs can be perceived as promoting anticompetitive behavior because they require integration and alignment between independent providers, as well as "consolidated" contracting.

For CINs to qualify as legitimately clinically integrated and thus be granted an antitrust exception, the FTC has outlined the following conditions:

+ A **network of physicians** willing to demonstrate **"a high degree of interdependence and cooperation,"** through
+ A **program of initiatives** designed to **"control costs and ensure quality,"** which
+ Is supported by an **infrastructure** that allows the physicians to **"evaluate and modify practice patterns."'**

In the Michigan area, there are two health systems that are collaborating to create what's called a "super-CIN," or a clinically integrated network that spans a region. The Together Health Network is led by physicians and powered by two healthcare systems, CHE Trinity Health and Ascension Health. These health systems have formed a clinical integration partnership that will encompass 27 hospitals and more than 5,000 physicians at hundreds of clinical sites stretching across the state of Michigan. The CIN has the potential to reach three-quarters of Michigan's patients while cutting costs and increasing efficiencies for the providers involved. Seventy-five percent of Michigan residents will be within 20 minutes of a Together Health Network provider. To be effective, they have to align their technologies, clinical pathways, and outcomes, as well as their electronic health records. When any patient in Michigan walks into any hospital, every provider will know that patient's medical history of procedures, medications, and other relevant health information. The intent is to reduce costs, increase efficiencies, and improve outcomes. Many organized health systems recognize the need to effectively manage the patient population as a whole. If they are unable to do it on their own, they are going to seek out partners, including their competitors, to do it.

ACOS: TRANSITION FROM ACO1.0 TO ACO2.0

Over 200 organizations participated in the initial Medicare Shared Savings Program (MSSP) to achieve the Triple Aim by establishing Accountable Care Organizations.

	Number of ACOs	Percent of Total
ACOs Generating Savings	53 (of 204)	26%
- ACOs with Hospital Participant	22 (of 85)	26%
ACOs Earning Shared Savings Distributions	49	24%
ACOs Generating Shared Savings that did not Earn Distributions	4	2%
Broke even	150	74%
Owe Losses	1	0.5%
Average generated savings	12.3 M	
Average earned shared savings payments	6.14 M	

In September 2014, when CMS announced the Medicare Shared Savings Program Year 1 results for 204 ACOs, only one in four systems demonstrated a savings. Overall, there were notable savings, but for some systems of care, the ACO model did not work well, largely because the focus was primarily on population health and chronic disease management.

While population health management remains a key objective in the current era, ACOs are now also focusing on specific therapeutic area initiatives and creating value-based relationships with stakeholders. Systems of care have begun working to establish payment schemes based on outcomes data, enabling coordinated care across multiple care sites and situations, and effectively empowering consumers to take control of their health and wellness to deliver a more holistic approach to care.

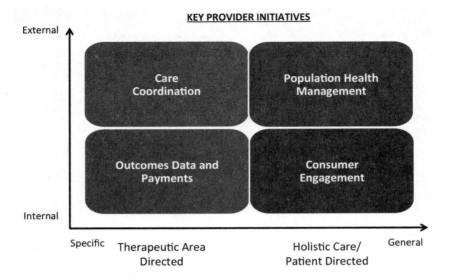

KEY PROVIDER INITIATIVES

Systems are at varying levels of sophistication and have differing financial situations, affecting which of these initiatives are prioritized and to what degree change is implemented.

Despite the lukewarm results from the MSSP, the number of ACOs is on the rise. The rate of penetration has varied significantly by state, with Massachusetts and Oregon being the two states with the highest number of ACOs and a large percentage of covered lives.

Systems of care have learned from this first wave, or what we will call ACO1.0, and are now expanding their focus. ACO2.0 is focused more on bundled payments for high-cost acute care and integrated care coordination. While ACOs do not represent all value-based contracts, they represent the most common response to commercial and government-initiated value-based contracts (VBCs).

ACO 2.0 requires different levels of care coordination, physician participation and governance

	ACO 1.0 Mixed results in the past 5 years based on a generalized approach to risk	**ACO 2.0** Approach to risk, will begin to evolve to the highest risk episodes for a provider based on their patient population
Clinical Integration	Care management for chronic populations, select acute	. . . plus additional acute and post-acute populations in narrow networks
Care Coordination	Evidence-based practices in ambulatory environment, pharmacy, lab	. . . plus acute and post-acute environment
Physician Participation	Primary care driven	. . . plus specialists, pharmacists, mental health professionals, and post-acute providers
Governance & Leadership	Physician leadership with business partners	. . . plus narrowed networks of allied health professionals, pharmacists, and post-acute providers willing and capable of sharing substantial risk
Payer Contracting	Primarily low risk around upside savings and FFS fee schedules	. . . assumption of upside and downside financial risk requiring more sophisticated actuarial analytics and predictive models to assess clinical and financial risk
Risk Mitigation	Provider credentialing and private inurement	. . . plus adherence to evidence-based practices, increased exposure to conflicts of interest
Capital Spend	Providers invested $2-5M to capitalize a panel of 5-20K patients	. . . plus operating costs to achieve scale and additional capital to assume risk

Source: Navigant ACO 2.0 briefing

Growth in ACOs within certain markets indicate providers' willingness to take on additional financial risk through new VBCs, and therefore ACO contracts can serve as a proxy for adoption of value-based initiatives.

VALUE-BASED PAYMENTS

To reduce costs and increase improvements in quality, providers are embarking on value-based contracts, which determine the distribution of risk between payers and providers, based on the patient population, and drive the potential payout from whatever savings can be achieved. To succeed in a value-based contracting world, systems of care need to be able to

> ✦ effectively structure a fair and transparent value-based contract, defining how any earned savings are distributed among providers involved in the ACO agreement

◆ analyze and uncover customer insights to understand behaviors and drivers/influencers of behavior of physicians and patients in the network

◆ determine the highest cost/highest risk populations and customize their approach at the specialty level

◆ collect, analyze, integrate, and share data to inform decisions among physicians, patients, and payers

The complexity required to execute at this level is significant and is outside the current capabilities of many systems of care. They are seeking partnerships in areas not seen previously and offering programs heretofore unheard of. Some progressive pharma companies are starting to explore partnership opportunities with providers to facilitate the transition to a value-based payment model.

THE IMPACT ON PHARMACEUTICAL COMPANIES

As we all know, the traditional pharma model involved personal interactions through sales representatives, medical liaisons, and key account managers, supported by nonpersonal marketing efforts. For patients, disease management and patient service hubs are established to support them and their caregivers. To gain formulary access, account managers would be deployed to work with each organization to understand the decision makers' formulary process and the stakeholders. Field commercial teams would focus on building awareness, providing education, and ensuring product availability; the field medical liaison team would reactively and appropriately address clinical and scientific questions posed by healthcare providers about company products that can only be addressed by a medical professional.

From a pharma perspective, what is the significance of the emergence of a super-CIN, where you have standardized care pathways and standardized order sets? Who now makes the decisions across these previously distinct organizations? Is that decision-making process slower or faster than it used to be? If you are on—or locked out of—one system's formulary, does that apply to the formularies of all the systems?

Pharma is beginning to ask these questions and adjust to the new reality of *who* their customers are and *how* they can best engage them. As noted above, key areas where providers need skilled partners include patient engagement, population health management, care coordination, and shifting to a value-based financial model.

There's a lot more at risk for pharma companies as these organizations join together and leverage their capabilities to not only deliver care but also consolidate their ability to evaluate and procure products, but there's a lot more opportunity as well. Therefore, the field teams (commercial or sales field forces, account managers, and medical field teams) should have a clear understanding of their customer(s) to better provide solutions for the systems of care and the care of their patients. A field organization will need to be more customer centric, proactively addressing customers' key questions (both from a clinical and value perspective) to show the economic benefit, the outcome benefit, other cost offsets, and so on. Healthcare providers are now being asked to justify their product/treatment choices, and so now they want to understand the clinical and economic value of products to help them support their decisions.

As pharma moves forward in encouraging these customers to explore new ways to strategically partner, it will be important for each company to ensure the engagement is done in a manner compliant with the respective regulations and the respective regulations, policies and procedures in place at each individual organization.

ADDRESSING ALL THE STAKEHOLDERS

Pharmaceutical manufacturers are already developing new approaches to engage with customer stakeholders, both traditional customers (physicians, nurses, patients, payers) and new or evolving customers (allied health providers, pharmacists, risk managers, administrators, care coordinators) who are becoming more involved in the delivery of care and decisions that may affect the utilization of pharmaceutical products and services.

Armies of sales representatives did the job successfully for years, delivering consistent messages to prescribers based on formulaic, predictive models. Primary care sales forces would call on primary care physicians, and specialty forces would focus their efforts on the specialists, and this model, for the most part, worked.

However, this approach has now been challenged, with access to providers, hospitals and physicians becoming more challenging. Some providers have instituted very strict "no-see" policies and policies that address the use of materials and products provided by pharmaceutical companies. The University of Michigan Health System was one of the first to limit interactions of their clinicians and manufacturers' representatives, and it included eliminating the use of drug samples,and instead, implementing a voucher system.

Pricing for new products has become much more scrutinized. To gain access to the market, products need to show differentiated/incremental benefit to the various stakeholders involved in the decision-making process. If there is no clinical benefit, no quality of life benefit, no improved patient services, or other compelling reason to gain access, the product must demonstrate a price benefit.

These changes are bound to have implications for pharma's sales force, its structure, composition, and capabilities. The traditional structure and activities of the sales force no longer apply. Most organizations recognize this but struggle with how the commercial team should evolve.

On one hand, they are dealing with highly innovative health systems such as the Cleveland Clinic, a cutting-edge organization providing holistic and accountable care. On the other hand, they are managing a number of smaller, less sophisticated hospitals and practices that are just implementing electronic health records or have yet to do so.

One of the key challenges for pharma is how to structure and address customer needs across the spectrum of the provider system's evolution from a fee-for-service healthcare model to a value-based model and how to best drive value for the various stakeholders. A pharmaceutical company must know where a system of care stands in its growth and integration (both horizontally and vertically) in the respective market. Where is the organization as it progresses from its largely fee-for-service environment (volume-based care delivered in silos, little incentive to integrate), which we'll refer to as "Curve 1," to an environment where, at the highest level of integration, it achieves the Triple Aim (better patient experience, improved population health, and lower per capita costs), which we'll refer to as "Curve 2."

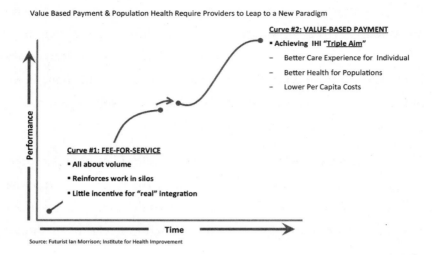

Value Based Payment & Population Health Require Providers to Leap to a New Paradigm

Curve #2: VALUE-BASED PAYMENT
- Achieving IHI "Triple Aim"
  - Better Care Experience for Individual
  - Better Health for Populations
  - Lower Per Capita Costs

Curve #1: FEE-FOR-SERVICE
- All about volume
- Reinforces work in silos
- Little incentive for "real" integration

Performance

Time

Source: Futurist Ian Morrison; Institute for Health Improvement

"Curve 1" systems involve traditional reimbursement models that put the majority of risk on the payer and incentivize transactional care, while the new "Curve 2" model increases provider coordination and integration with a focus on quality and lower costs.

Each system will have different needs and objectives, and a one-solution-fits-all approach by pharmaceutical manufacturers will be challenging. The approach to Geisinger Health System (one of the most successful highly integrated systems in the USA), for example, will need to be different from the approach to a less integrated system. Pharmaceutical companies will need to identify their key customers and understand and track where these customers are today and where they intend to move to in the future. This will enable the development of strategies and potential solutions based on the evolution of those customers. Solutions may include "beyond the pill" (or value-added) services that need to be considered to further differentiate the product, portfolio, or company offerings to help drive value. Some of these services may be product or disease specific while others may be more focused on helping providers manage their populations more effectively.

An example of a collaborative effort that includes the pharma industry is the collaboration between the Center for Connected Health, a division of Partners HealthCare, and Daiichi Sankyo to create a mobile app that will serve as a coaching platform for patients with atrial fibrillation who have been prescribed oral anticoagulation therapy. The goal of this app will be to support patients with atrial fibrillation by helping improve patient adherence to and compliance with their medication regimen. The app will also foster feedback loops that connect the provider to the patient.

And there are other stakeholders who must not be overlooked including pharmacists, nurse practitioners, and physician assistants. Pharmacists in the retail setting, historically, have been important healthcare providers, providing product information and advice to patients. But they are becoming more important and stronger stakeholders because there is a push to get patients into settings where the cost of care is lower. Pharmacists, therefore, are taking on a much bigger role in helping patients with their therapies, answering questions not only about the medication (dosing guidance, side effects, etc.) but also about costs and, in many cases, about alternatives if the out-of-pocket costs are too high for patients, and they are also providing ancillary services. This is another group of stakeholders whom pharma should pursue to develop a strong relationship and provide appropriate support services.

PROVIDERS TAKING ON RISK

ChenMed, founded by Dr. James Chen in Miami, Florida, absorbs the risk of patient management by taking responsibility for getting chronically ill, senior patients to their outpatient appointments. Primary care compliance through programs like ChenMed is targeted to the most vulnerable patient population: moderate to low-income seniors with complex chronic diseases. It is 100 percent risk bearing and treats its patients within a closed system. The ChenMed model, which has been expanded beyond Miami as JenCare, has shown a 38 percent reduction of patient days in hospital in audited 2011 data.

Providers, like ChenMed, are addressing patient noncompliance and/or poor adherence that can be very costly to government and commercial payers.

ChenMed members who pay for their care services are obligated to come in for a monthly appointment. The organization utilizes its own vans and cars marked with the ChenMed brand to pick up the patients and bring them to their appointment. The high-touch relationship between ChenMed and its patients not only reinforces the desired healthy behaviors but also leads to higher patient satisfaction.

The push to get patients into lower-cost settings also has created a greater reliance on nurse practitioners and physician assistants in managing patients. Utilizing these providers reduces the cost of care and helps free up more costly resources for the benefit of higher-risk patients. This is an audience that will likely have greater influence over time. Nurse practitioners and physician assistants either influence or directly prescribe medications (governed by state regulations based on the level of physician involvement). Pharma will need to continue to cultivate and provide specific customized support services or programs for nurses and physician assistants.

Because of healthcare reform, other key stakeholders are emerging who must be taken into account. They include quality managers and risk managers. The integration and risk associated with managing populations and the reimbursement implications mean health systems must increase their focus on delivering quality care and producing quality and performance metrics. These stakeholders are necessarily engaged in developing a robust quality and performance improvement infrastructure using sophisticated data analytics and aligning physicians to achieve the organization's objectives based on Triple Aim goals.

PARTNERING WITH DATA

The availability of electronic medical records and the wealth of information available in this era of big data do not only have an impact on the delivery of care but also on how product interventions are measured. Although the use of data is nascent but growing rapidly, it can provide tremendous insights on how best to use new therapies or technologies to improve quality care.

This, in particular, is where pharma can bring a lot of value. At present, there are few examples of organized systems of care using this data to provide important insights. They have not yet developed the analytical frameworks or the technical support to be able to integrate this data.

Systems of care are starting to reach out to pharma manufacturers for assistance in standardization of care. Much research has indicated that the variation of treatment within an institution is a primary contributor of high costs per episode. Systems of care are seeking to address such variation through chart/claims analyses, as well as driving physician alignment around the need for standardization.

The pharma industry has long developed the infrastructure and analytical frameworks to generate important insights about their products and customers. A core competency of the pharmaceutical industry is the insightful analysis of large amounts of data, which the industry uses to further its progress. This is a unique capability that pharma can bring to the health systems with which it strategically partners. The pharma industry has been doing analytics for years to understand trends, gain insights into physicians' practices, and understand patients' reactions to, and behavior around, treatment therapies. Through analytics, the industry has uncovered greater insights into the market and into increasing the use of pharmaceutical products.

The need for deeper analytics to support treatment management may present an opportunity for the pharma industry and health systems to come together and start sharing information and data. If pharma companies can identify regulation-compliant ways to partner with key accounts to use data to evaluate specific patient populations, they may create a win-win situation for themselves and their customers.

Pharma could benefit from the standardized use of its drugs in the most appropriate patient types. Providers/payers could benefit from the reduced churn of patients, thereby facilitating patients' access to an appropriate therapy.

In June 2014 Cigna and AstraZeneca engaged in the analytics of medical and pharmacy claims data to develop predictive risk modeling that would assess a patient's risk of atherosclerotic cardiovascular disease and the impact of managing the patient's cholesterol. The objective was to understand which patients were at the highest risk so the right therapies could be targeted to the high-risk populations through appropriate controls that would reduce low-risk patients' use of high-cost medical therapies.

Improving patients' adherence to their medication regimens is one important example. The issue is not a simple one to address. Data indicates that after three months, about half of the patients on medications for chronic illnesses, such as hypertension, high cholesterol, and diabetes, discontinue therapy. After a year, only about 30 percent still take their medication, thereby creating a situation in which patients' health deteriorates and the risk of developing complications increases. Improving medication adherence provides both short-term and long-term benefits for both pharmaceutical companies and health systems managing population risk.

Adherence is challenging for a variety of reasons that include patient cost concerns, the burden of taking multiple medications (for patients with multiple comorbidities, such as hypertension, diabetes, and cardiovascular disease), side effects or tolerability concerns, and just plain forgetfulness. Pharma can provide an analytical powerhouse to identify and engage patients to better address their specific needs, attitudes, and behaviors.

For example, Geisinger and Merck have a multiyear collaboration on health IT programs that targets improving patient adherence to treatment plans and clinical care processes. Most health systems do not have the capabilities to develop tools for patient risk stratification. As part of the collaboration, an interactive web application was designed to help primary care clinicians assess and engage patients at risk for cardiovascular disease and type 2 diabetes. The innovation will be tested and refined in the Geisinger network. The objective is to address key needs and issues regarding patient care, including failure to adhere to medication—which can result in readmission— as well as effectively stratifying patients based on their risk level to better identify and manage the most costly, at-risk patients.

Pharma can also use its sophisticated analytical capabilities to collaborate with health plans to determine which interventions should be developed to address patient needs and behaviors. These interventions will be tailored to systems, patient types (segments), geography, and market location.

The health system wins when patients remain on their therapeutic regimens, as these patients will hopefully stay healthier longer, reducing the likelihood of hospitalization or other high-cost care. Pharma companies win because the patients make full use of the products for the long term, and the medications deliver on their promise to enhance the health of the patients. This ensures a positive experience for physicians and patients, creates greater revenues for the manufacturer, and provides additional data regarding the effective longer-term use of medications. This data can be further analyzed to enhance the value proposition of these products for all stakeholders and can also provide greater insights into how both organizations can work together to improve or reconstruct effective programs and interventions.

VALUE-BASED CONTRACTING

The number of payer-provider relationships may increase if providers become more comfortable with accepting financial risk for patient outcomes and can develop the capabilities required to implement and manage population health programs. Because most organizations don't have the capabilities required, payers and providers who are at risk have been reaching out to pharma and medical device companies to form partnerships. Most partnerships we've seen thus far have focused on developing disease-specific tools for patient adherence or care coordination, while only a few arrangements have surfaced in which the patient outcome puts a pharma or device supplier at financial risk.

The most cited example of such a partnership is the trend-setting Merck and Cigna contract of April 2009 that was designed to increase patient adherence among Cigna's type 2 diabetes patients. Medical costs for people with diabetes are 2.3 times higher than the general patient population, and poor adherence to therapy strongly correlates to increased utilization and cost. Through the execution of a value-based contract with Cigna, Merck assumed responsibility for improving diabetes medication adherence. This deal was highly innovative when launched in April 2009. Merck contracted with Cigna to decrease the price for its branded diabetes drugs Januvia (sitagliptin) and Janumet (sitagliptin/metformin) and offered additional discounts via rebates to CIGNA when its members with type 2 diabetes consistently achieved guidelines and recommended levels of HbA1C and adhered to their prescribed therapy. Merck offered a second round of rebates when patients lowered their blood sugar below recommended guidelines. By the end of 2010, Cigna had achieved more than 5 percent blood sugar level improvement for those 165,000 members continuously enrolled in the program using the Merck diabetes drugs, and drug adherence improved to 87 percent.

Among medical device manufacturers, the use of value-based contracting has higher impact with payers and providers because device use may be directly tied to high-cost surgical time and use of hospital beds. St. Jude Medical is one of many medical-device companies that announced risk-sharing agreements with providers in 2014. St. Jude recognized that heart failure is a critical quality metric for most of its customers, and poor device implantation can lead to increased costs/decreased reimbursement. St. Jude Medical put itself at risk for 45 percent of the net price of a cardiac resynchronization therapy (CRT) product if a revision surgery of their implantable CRT was required within one year of initial surgery. St. Jude's value-based guarantee is narrow but is seen as a step toward aligning payer, provider, and supplier incentives.

It is important for any company considering value-based agreements to note that each company will have its own compliance and legal standards concerning the level of risk it is willing to accept. Company compliance officers and legal departments should be closely engaged in the design of such agreements since they can put the organization at significant financial risk.

LEVERAGING THE NEW OPPORTUNITIES

The changes happening in the healthcare market are creating new opportunities for pharma manufactures, including:

+ **Supporting patient or consumer engagement in healthcare.** This goes beyond pharmaceutical companies directly engaging with patients or consumers (through direct-to-consumer models or product education). Rather, it helps healthcare providers and health systems to engage consumers and patients effectively to drive optimal care. Pharma companies have tremendous insights into communicating information to patients, based on their history and competency in direct-to-consumer modeling. By using this knowledge and capability, pharma may be able to support system-of-care goals for patient engagement to further enhance care.

Boehringer Ingelheim (BI) has partnered with ApolloMed ACO on chronic obstructive pulmonary disease (COPD) management to promote best practices. This collaboration provides an example of patient engagement with the intent to improve performance and quality. Under this program, BI offers resources from its "Strategies for Chronic Care" and "InStep" programs to ApolloMed ACO to support the management of patients with COPD. In the arrangement, BI provides training and evidence-based protocols. There is no risk sharing between the parties. However, the program tackles key needs including a knowledge gap regarding patient management best practices and patients' adherence to their care plan. Patient adherence has an impact on quality measures because COPD is a leading cause of readmission to the hospital and in 2015, a CMS quality measure for readmissions of adult patients with COPD will have an impact on Medicare reimbursement.

✦ **Supporting the comprehensive care coordination model.** This model requires a highly coordinated delivery of care to patients that includes providing the right level of care at the right time in the most appropriate location by the right provider. Pharma has tremendous expertise on understanding the patient journey and how to identify the leakage points where patients fall out of the system. Pharma companies can leverage this expertise to provide enormous insights on patient treatment flow models for providers, including opportunities on where to engage patients or improve care. Pharma can provide the expertise to develop outreach programs and interventions during the patient journey at points where, for example, they are likely to fall off a medication or therapy or interact with a healthcare provider (e.g., the patient's visit to a pharmacist to get a prescription filled provides an opportunity for the pharmacist to deliver messages reinforcing the right patient behavior). By collaborating and using these insights and information, organizations can determine how best to provide coordinated and comprehensive care and keep these patients in the system.

Merck spin-off Vree Health collaborates with Frontier Medicine Better Health Partnership (FMBHP) to improve care transitions for patients in rural areas. Under the agreement, FMBHP uses Vree Health's service, TransitionAdvantage™, as a pilot to improve care transitions from hospital to home and reduce hospital readmissions for patients in rural areas of Montana. The program addresses an unmet need in the continuum of care for rural providers, as populations in rural areas have difficulty engaging in post-discharge care and regular follow-up visits. Readmission rates in rural areas have been particularly challenging due to poor patient follow-up and care post-discharge. Following through on care transition is challenging in large catchment rural areas due to transportation time.

✦ **Providing healthcare provider organizations with the right clinical or performance endpoint data for their products or services.** Study endpoints should align with the incentives or goals of the healthcare system or provider groups. Consideration should be given to the development of novel endpoints. These endpoints would go beyond those needed for regulatory approvals or even traditional endpoints that provide comparison versus placebo or a comparator or improvements from baseline. Novel endpoints could provide insights into the ability of a therapy or technology to provide increased care to patients and potentially help systems of care deliver on quality and performance targets. Some companies could even consider partnering with health systems to explore the development of novel endpoints to validate their importance and relevance.

✦ **Enhancing the patient experience.** This has taken a number of forms, including a greeter for patients entering a facility, a coordinator to help patients move and navigate the health system, and patient satisfaction surveys. Sometimes, the need to enhance patient satisfaction encounters opposing forces. Imaging centers, for example, are encouraged to enhance throughput (drive greater volumes of patients through the system as fast as possible) to increase cost effectiveness. However, this increased throughput is expected to occur with existing resources, minimal wait times, and high levels of patient satisfaction.

Improvement of patient satisfaction is particularly scrutinized because patient satisfaction survey scores are tied to the provider reimbursement rates. This may well be another area where partnering or value-added services can be effective.

The US healthcare system is undergoing significant change through a variety of reforms to optimize the delivery of care clinically and economically. These reforms are intended to address the cost concerns of the most expensive healthcare system in the world, as well as improve the management of acute care episodes. They are also meant to address issues surrounding chronic conditions that are highly prevalent and drive increased use of the healthcare system. Numerous initiatives and pilots are being pursued to determine the optimal way to deliver care in this market. These have included shifts in incentive systems, performance-based models, the emergence of new stakeholders, and the shifting of risk and cost from employers and payers to providers and patients. There has been unprecedented integration and consolidation among providers to form networks that allow the necessary scale to compete effectively.

These changes have caused the pharmaceutical industry to consider how to best reconfigure its relationship with health systems. Pharma may be uniquely positioned to support the challenges of these health systems. It will be critically important for pharma to understand how to structure and address customer needs based on where a health system is in its evolution from a fee-for-service ("Curve 1") healthcare model to a value-based model ("Curve 2"). Initially, the most interesting and achievable areas for potential collaboration include patient/consumer engagement, care coordination, patient satisfaction, and potentially value-based contracting. Pharma can strategically engage with these customers, leveraging core capabilities (including data analytics, in-depth knowledge of patients and physicians) to address these important challenges and forge new partnerships to deliver the best therapies for patients.

CHAPTER 5

THE PHARMACEUTICAL MANUFACTURERS' PERSPECTIVE

IN PREPARING THIS BOOK, IN ADDITION TO ALL OUR REGULAR ENGAGEMENTS WITH CLIENTS ACROSS THE INDUSTRY, WE SPECIFICALLY INTERVIEWED DOZENS OF EXECUTIVES IN PHARMACEUTICAL AND HEALTHCARE MARKETING. SOME WENT ON RECORD; OTHERS ASKED TO BE ANONYMOUS. THEIR THOUGHTS, IDEAS, COMMENTS, AND PERSPECTIVES ARE SUMMARIZED THROUGHOUT THIS BOOK AND IN THE PAGES THAT FOLLOW.

In writing this book, we wanted to go straight to the front lines and talk with people who are working on these challenges and shaping the future of pharmaceutical and healthcare marketing today. We wanted to understand their views, concerns and challenges, and the evolving opportunities they saw on which marketers of the future could capitalize.

The opinions and statements expressed in this book represent the personal opinions and views of the individual contributors cited, not necessarily the views of their respective companies or past or present employers.

Even with such as disclaimer, the regulatory environment is such that some individuals still requested to be anonymous.

We interviewed many people for this book and used direct quotes from the following people regarding the state of the industry and the challenges, trends, and opportunities that lie ahead. In the pages that follow, we have done our best to summarize and include the consensus of all such views.

Pat Andrews, executive vice president and chief commercial officer at Boston Biomedical started her pharmaceutical career at Pfizer almost 25 years ago. After becoming the vice president and general manager of Pfizer's USA oncology business, Pat left to build the commercial organization at Incyte and launch its first-in-class, first-in-disease, first-for-the-company oncology product.

Tim Cole has over 30 years' experience in the pharmaceutical industry. While Tim had responsibility for both sales and marketing for part of his career, he is, essentially, a sales executive, which is his current position and accounts for most of his experience.

Craig DeLarge has worked in many senior level digital healthcare roles at J&J, GSK, Novo Nordisk and Merck. He is currently Digital Healthcare Strategist at WiseWorking, LLC, a digital healthcare and change leadership practice.

Susan Harris started her career as a consumer package goods marketer at Procter and Gamble. After about ten years of package goods experience, Susan joined Pfizer to continue her marketing career in the pharmaceutical industry. She is currently a marketing vice president at Novo Nordisk with over 20 years of prior experience at companies such as Wyeth/Pfizer, BMS, and Sanofi.

Yolanda Johnson-Moton is director, external relations, US medical affairs for Eli Lilly & Company. She has held several roles around strategy and influencing customer perceptions in her seven years at Lilly. Prior to Lilly, she held various positions at Merck in the cardiovascular franchise.

Peter Justason has worked in many senior level digital marketing roles at both Johnson & Johnson and Purdue. He currently serves as the director of emarketing at Purdue.

Lars Merk is an 18-year industry veteran who has held numerous roles with Johnson & Johnson and its subsidiaries. His background includes working in and managing field sales forces, e-learning, product roles in gastroenterology, pediatrics, neurology and CNS, and digital channel responsibilities for numerous OTC brands. He recently joined AstraZeneca as portfolio marketing director of their diabetes franchise.

Shawn O'Hagan has worked in project management, senior-level marketing, and digital marketing roles within the pharmaceutical industry for the past fifteen years. His current role is within multichannel marketing at Daiichi Sankyo.

Catherine Owen began her pharmaceutical career in the UK with Astra Zeneca, and after just a few years, joined Johnson and Johnson. She spent over ten years with J&J in the UK, eventually moving to the USA where she has held many positions with increasing responsibility. Her current role is that of vice president of immunology marketing at Janssen Biotech Inc. of Johnson & Johnson.

Scott Richardson is vice president of global marketing at Pfizer and has been with Pfizer for over ten years. Prior to Pfizer, Scott spent 13 years at Cigna, eventually becoming a vice president of product strategy at Cigna. Scott brings his experience on both the pharmaceutical manufacturer and managed care payer sides of the business to work every day.

Arnel Rillo is founder & CEO of eVeritas, a clinical solutions software platform that spans the entire continuum of care for chronic disease management. Prior to that, he was senior director of disease management at OptumHealth, a division of United Health Group, and before that he spent almost 20 years at GlaxoSmithKline and its predecessor companies in a variety of sales and marketing roles.

John Russo is Senior Director of Marketing at Ethicon. In his more than 14 years at Johnson & Johnson, he has had many roles in sales, marketing, and strategy and has been responsible for both pharmaceutical and device brands.

Sanjiv Sharma is truly a global industry executive with over 30 years of experience in India, Canada, and the USA. He has worked at large companies (Sanofi) as well as start-up pharmaceutical companies in positions in sales, strategy and analytics, marketing, and general management. Recently he was founding principal of InflectionPoint, a strategic consulting company. Prior to Inflection Point, Sanjiv successfully launched Duchesnay USA as head of commercial operations.

INDUSTRY TRENDS

In the years ahead, every pharmaceutical company that remains in business will likely take one of the following four paths forward or a combination, which we are seeing a lot as companies decide their future.

- ✦ Stay with a predominantly manufacturing model, such as a generics company.
- ✦ Become a specialty pharmaceutical company focused on biologics, rare diseases, personalized medicine, and so on.
- ✦ Move up the value chain and become an outcomes-focused healthcare company by moving "beyond the pill" to combine therapeutics with information and social services.
- ✦ Merge with another organization.

Initially, most are not making "all-in" bets but have already begun experimenting with these options. Specialty-focused niche drug development is where we are today because it currently represents very high margin opportunities. Most are complicated and expensive medicines, and even though they have smaller patient populations, the profit margins on these drugs are reminiscent of days gone by. It's a very good business if you can charge tens of thousands of dollars for a hemophilia or hepatitis drug because you have true product differentiation or are the only option available, as long as there are willing payers.

These drugs are no less expensive to research, develop, and manufacture, and companies still have to ensure the safety of the supply chain. In these respects, they cost no less to produce than the blockbuster drugs of the past, but they only treat a small fraction of the population. Some rare diseases affect just a few thousand people a year. To help explain why they cost as much as they do, imagine how much an iPhone would cost if the total market of potential buyers was less than 10,000 people. Would Apple even invest the R&D? Probably not, because iPhones don't save lives, and people would not value the phone enough to pay what Apple would need to charge to make it profitable. At the end of the day, with quality specialty drugs that uniquely save lives, the margins are going to be fairly high as compared with lower margins for a pill in a class with plenty of "me too's" or generic options. The economic benefits—at the organizational level and in payer negotiations—of concentrating on core therapeutic areas are becoming increasingly important, as evidenced by the recent asset swaps between Lilly, GSK, and Novartis.

"Virtually every pharmaceutical company has an innovations group or small team, or at least an executive who is focused on 'beyond-the-pill' initiatives," says Peter Justason.

Merck has been moving beyond the pill in many ways as it follows an acquisition and investment path and either buys or invests like a venture capitalist in many digital healthcare companies. "I give Merck credit for investing aggressively in various digital healthcare companies. But I can't help but think about the example of the Sony Betamax and the VHF. Placing bets on single entities is a dangerous process," says Lars Merk.

GOVERNMENT'S INCREASING ROLE

"A few years out, I don't get paid by the government unless I can show, as an accountable care entity, that this group of patients in this geography with this particular health problem have measurably improved in terms of their health outcomes. That is the result of an aggregate solution offering that includes therapeutics, behavioral modification, social and emotional support, information and counseling. Think of that as 'the product'. And as an outcomes company, to get paid, the company will need to implement such a sweep of solutions with a given population," says Craig DeLarge.

You can view a great video interview with DeLarge that expands further on these thoughts as well as implications for the publishing and support services companies within healthcare: http://www.ehealthcaresolutions.com/interview-craig-delarge/

Tim Cole from Sanofi has a similar perspective. "When I talk to physicians associated with an ACO and patient center medical homes," he says, "they are starting to think about more than efficacy, safety, tolerability and cost to the patient; they are starting to think about how they will get paid by the government."

Sanjiv Sharma puts it simply: "The Affordable Care Act is changing the game completely."

"The United States is a free market that doesn't enforce price requirements. Historically, industry has been allowed to price at whatever it chooses and increase price at will. Over the next five years we are moving toward socialized medicine similar to NICE (National Institute for Health and Care Excellence) in Europe, and the federal programs in Canada (Canada's Patented Medicine Prices Review Board). They don't see safety and efficacy and the economic benefits as two separate discussions. Drugs have to be remarkably better than the standard of care, or they don't pay the premium. Going forward, to get prices approved, industry will have to prove, through pharmacoeconomic studies, clinical trial data, and increasingly through ongoing real-world outcome studies, why a premium price over existing medications is justified," says Peter Justason.

Government regulations also have made it more difficult to market our products. Pat Andrews makes an analogy to the computer industry: "Several years ago, Apple was running a TV ad where a young hip guy was standing next to a more conservative guy who represented Microsoft/Windows. The Microsoft guy was counting money and putting it into piles, one huge pile for advertising and one small pile for R&D. The Apple guy had the exact opposite situation, lots of money into R&D and little money into advertising." Pat says, "That is the way leading pharmaceutical companies have been successful—we put money into R&D and created new and better products. We are a highly regulated industry; we can't just advertise we have a better product, we truly need to make better products. We need major steps forward, not incremental change."

PAYERS

"There used to be one word for *payer*. Now we refer to market access, managed care, provider networks, insurance companies, integrated delivery networks (IDNs), accountable care organizations, regional health information organizations, the government, and various others groups all as 'payers'. There is a level of complexity and granularity as each organization holds slightly different roles in a fragmented healthcare system, and their scale is growing. We have to deliver big-scale programs to payers that will improve outcomes. These programs have to save them at least a million dollars or more to makes it onto their radar," says Peter Justason.

Each group has different nuanced needs and motivations. Each must be addressed. Each healthcare market, pharma company, medical device and diagnostics (MD&D) company, lab testing firm, hospital, healthcare provider, and so on, faces its own unique challenges with payers.

"Payers are changing, and healthcare providers' worlds are being disrupted," says Lars Merk. "Their compensation, motivations, and capabilities to be sole proprietors of their own practice are all decreasing. The rise of accountable care organizations feels like history repeating itself. The new principles we're seeing, both good and bad, are recycled variations from the 1980s and the rise of HMOs," says Lars Merk.

"Lowering spending on healthcare is not the goal. The goals are to improve our health in the overall population as measured by quality of life years and to give more people access to healthcare and to medications. Patients don't want to pay more, but they will. Pharmaceutical companies don't want to discount more, but they will. Payers don't want to pay more, but they will. And healthcare providers don't want to give up reimbursements, but they're going to have to do so. Nobody is happy with it", Merk says, "but it will work."

PHYSICIANS' DIMINISHING ROLE

As physicians become employees of larger conglomerates, decision making is being taken out of their hands. Prescribers are inundated with patients and the increasing requirements of their workday. Managed care is top of mind not only for prescribers—since it is taking away their choice and influence—but for patients and pharmaceutical companies as well. A big shift is coming that will drive marketing efforts more toward the patient and the payer.

"Empowerment or lack of…physicians have less influence over the prescription choice, due to strict ACO, IDN, and managed care guidelines," says Shawn O'Hagan.

Today, a pharmacy and therapeutics committee, a medical director, or a group of influencers within an integrated delivery network make decisions about which drugs within a certain category or class should be on formulary. By the time their decision trickles down to physicians, it has become two or three choices on a pick list on their handheld device. As pharmaceutical companies come to grips with managed care's influence, they are continuing to scale back their huge investments in sales organizations and sales infrastructure. Sales representatives are simply becoming increasingly ineffective at getting access to the new decision makers.

CUSTOMERS DEFINED

So who is the customer of the future? Some would say it is clearly the patients, the ones who put their ultimate trust in us by consuming our products. Others would argue it is the payers. After all, they are the ones paying for the product. Just follow the money. Most agree that both will continue to be incredibly important in the future.

"For the most part, customers can be summed up as the four Ps: Patient, Prescriber, Payer, and Pharmacist. However, the new entries are ACOs and IDNs. Regardless, each of these need to be communicated to in a more customer-centric way," says Shawn O'Hagan.

"Customers are changing. Gone are the days when the clinicians have unlimited choices. A financial buyer has defined in advance a limited subset of products available to them. This is perhaps even more impactful in the world of medical devices than it is in pharmaceuticals. With drugs, there will always be patient populations that don't respond to one medicine but do well on another, or populations that have side effects and require alternatives. There is a first-line therapy, second line, and so on. With medical devices, as the financial buyers make their decisions on standardization, the winners and losers will find they are playing an all-or-nothing, zero-sum game. This is a scary change because we haven't traditionally called on financial buyers. We don't have relationships with them, and we don't really understand them. For all of these reasons, digital advertising specifically targeting the nonclinical, financial buying customer is a powerful tool for MD&D," says John Russo.

PHARMA SALES FORCES

"Our access to physicians for our representatives is diminishing rapidly because individual doctors' practices are being purchased by IDNs and ACOs. The IDN or ACO then institutes a policy preventing physicians from spending time seeing pharmaceutical representatives. Our most effective channel, the sales representative, is effectively being blocked," says Peter Justason.

As with many other aspects of business, where a need exists on a shorter-term basis, the shift to outsourcing will likely continue. "There will be a radical increase in the contract sales organizations (CSOs) where sales representatives will actually get a little more power than they have today," says Lars Merk. "If they're selling to endocrinologists and have strong relationships in an area, companies will bid for their time to sell their products, versus the other way around where CSOs are trying to win over companies that have products. CSOs will own the relationship, just as they're doing this on the clinical research side today. I believe you will see 50 percent of the sales forces outsourced by 2020 in a 'virtual detailing' or 'just-in-time' detailing capacity."

Imagine a physician in the future who is looking at a patient's electronic medical record and wants to prescribe a statin, but the patient has a comorbid condition. Instead of searching for answers, the physician simply clicks a button that immediately opens up face time with an expert CSO-contracted trusted representative who can answer the questions if the label provides the answers or can immediately transfer the doctor to a company medical science liaison. This might be a live interaction, or a virtual answer that comes up in a search result if we know that the questions are being posed by a physician.

"Our reach right now with the nonpersonal promotion is about 95 percent of all the targets we have. So today I can reach more than our representatives can and that gap is just going to grow larger over time as no-see doctors increase," says Peter Justason.

"We actually do digital nonpersonal promotion to IDNs in the United States, and we have found out they are hungry for information. As a result, their engagement is two to three times above average. There will be a marketing mix with IDN physicians in which some are going to want e-details or virtual details, some will want to access our call center, some will engage in our online case studies, and others will just visit our websites. Unbranded activities that help them in their practices such as our patient support materials for chronic pain management are also valued. We download thousands of pain scales every month that are unbranded but serve as a useful resource these doctors want and use. The biggest opportunity with IDNs are digital initiatives supporting outcomes. We want to help them manage chronic pain and save them money. It's a win-win for all of us because it means better outcomes for patients and lower costs for the IDN," says Peter Justason.

Methods other than the sales force become more important as companies begin thinking about their portfolio of brands. Catherine Owen, a vice president at Janssen, makes it very clear: "Most people in pharmaceutical marketing have grown up working in businesses with a one-brand focus and having to worry about ensuring the sales force focus on that one brand. As we are beginning to have more portfolio roles, being responsible for multiple brands competing in the same market, this is a problem the industry needs to solve."

Scott Richardson from Pfizer thinks about portfolios and sales forces similarly. "We are treating a patient with a lot of comorbidities," he says. "So the question is do you position your portfolios in a different way that's meaningful to a patient and to the physician? Within pharma there are different lines of business—mature products, launch businesses, innovative businesses—so you need to think about a segment that you are trying to achieve based on the unique aspects of your products and balance your needs with the needs of the customer. We need to leverage these portfolios in a meaningful way to our customers that is going to keep them top of mind and add value to their lives. And we have to figure out how the sales force is integrated and can help with the issue."

Marketing in the medical device space is about education. Specifically, it's about teaching the surgeon how to use the product safely and correctly. "If I can train a surgeon to use my device safely and correctly via their mobile device," says John Russo, "it helps my representatives in the field and allows them to focus on driving trial."

SOCIAL MEDIA

One executive indicated that 70 percent of the site traffic on one of the company websites came from Facebook.

The excuse often cited for why a brand does not have a meaningful interactive social media presence is that obligations will arise if an adverse event is reported. For example, someone might post that a product is not working as specified, as in, "I took my Ambien and still couldn't get to sleep." At this point, there is still a lot we would need to know for that to qualify as a reportable event. We don't know if the patient took it correctly. The user is only identified by an anonymous handle, so we don't know his or her name. Perhaps, the user took something else, such as Lunesta, and is just saying "Ambien" to mean any sleep medication. But we should want to find out. This is a customer with a problem. In any other business, that spells opportunity.

We're overly worried about our reporting requirements and label changes that could come from the posting of an adverse event. If we truly care about people's health and we want to improve the product, we need to listen to customers on social media. Lars Merk suggests that we embrace these opportunities to find out what happened and use that information to better understand our products and improve them.

"The appearance of adverse events online is significantly lower than most drug safety groups fear. It is very manageable for a company to deal with. But drug safety groups are concerned at the possibility of the floodgates opening and being overwhelmed. The amount of actual qualified adverse event reports is tiny," says Peter Justason.

"Over the years, the industry has gotten itself into bad situations with many products only after patients have used them on a large scale. Where appropriate, social media could help us move products toward new uses and away from harmful uses faster. Social media can serve as a research window into real-world usage and outcomes on a large scale. Had it been available previously, the social media might have minimized or avoided some of the product liability suits that challenged the pharmaceutical industry. We may have known about issues sooner. Armed with this knowledge, we might have avoided some lawsuits, penalties, and judgments, and more importantly, we might have saved people's lives or protected their health. We have a responsibility and an obligation to listen," says Merk.

DIGITAL MEDIA SHIFT

"I think what we will see is a redeployment out of television and mass media into building services that are relevant to our customers. There is going to be a continued rise in the importance of the investment in digital marketing within the overall marketing mix of a product," says Peter Justason.

"Digital gives us the ability to laser-target. I can reach a physician by name multiple different ways during a single day. So we can laser-focus physicians with laser-focused messaging. Targeting gives us the ability to measure ROI, the ability to prove the ROI preserves the budget for the best performing tactics. As marketers, we want to deliver the best ROI for the company and shareholders, and that starts with sophisticated targeting and measurement. I think that by 2016 most budgets within pharma brands are going to be more than 50 percent invested in digital marketing. While I don't think print is going to die altogether, I do think it's going to be in a long-term-care facility going forward," says Peter Justason.

SHIFTING FROM PRODUCTS TO OUTCOMES-BASED SERVICES

"Imagine a brand.com site for a diabetes drug produced by a manufacturer that produces no cardiovascular drugs. The manufacturer knows, however, that cardiovascular disease is a common comorbidity with diabetes so that manufacturer provides information not in its own interests but to improve the patient outcome by assisting the practitioner or patient experience," says Lars Merk.

"This redefines 'coopetition' yet again. The pharmaceutical industry has to not only break down its internal silos but it also has to open itself up to collaborative external partnerships," continues Merk. "Why not facilitate a warm handoff to Weight Watchers or Jenny Craig to help patients with their weight management? Or to Fitbit, Moves, or Apple Health to help them get better control of their activity levels? Those are the types of discussions that we need to be having. Why are we putting healthy menu items on a diabetes medication website when we have no expertise in creating recipes? Why not partner with somebody who actually has a stake in that rather than pay someone else to do it for us?" asks Merk.

Sanjiv Sharma believes that "as the industry is moving toward smaller and smaller niche products, some of these niche products need a device to help diagnose that niche disease. Thus, complementary products are critical. You see lots of this in oncology, and as we continue to develop more niche products, this will become the norm."

Susan Harris is a big believer in outcomes versus products: "It is not just about the product, but it is about the service that wraps around the product that then delivers better outcomes. When you are delivering better outcomes, you can keep people out of the ER; you can keep people from getting complications; you can keep people from going back to the hospital for readmission. It is better for the patient and the physician. That's where I see the industry going, but we are struggling with it because we're rooted in a product model." Harris also believes that as the industry moves to a focus on smaller disease states, outcomes become more important, and this will greatly reduce marketing budgets and force the industry to get smarter.

MOBILE

The consumer electronics industry is going to continue to innovate at a faster pace than the healthcare industry. Companies such as Apple, either with or without pharma, will connect healthcare to the mobile capabilities of the phone: long battery life; geolocation; GPS; processing power; great visualization; ability to sense when other people you know are near; search; videos, and so on. Think about adapting each of those capabilities to healthcare. You can make medical devices a lot cheaper and a lot dumber, with the smartphone doing the heavy lifting.

"By 2015, at least 25 percent of pharmaceutical media is going to be mobile specific," says Peter Justason.

One powerful moment for a marketer to support a patient is when they are sitting in a physician's office, worrying about their condition, their test results, or awaiting a diagnosis. "This is one of the moments when people are really thinking about their healthcare. It is a great intervention point. We can prepare patients for their discussion with the physician and answer their questions immediately after the appointment or introduce them to a treatment option after a diagnosis. When we add geolocation mobile messaging, there is huge potential to interactively create value," says Merk.

Imagine if somebody walks into a CVS pharmacy and they have four prescriptions, if we could tie in their payer data and find they haven't filled a certain prescription in a prescribed amount of time, we could intervene and improve the outcome by ensuring they fill their prescription. Everybody thought e-prescribing was going to help with prescription fills, but now that the patient doesn't even have a piece of paper, even more prescriptions are being abandoned. This is something that has to change if we are to impact outcomes. Medicines work and they work cost-effectively," says Merk.

When the trend involves disruption of an industry, everything we do has to be completely re-examined. For example, Lars Merk asked and answered this question and gave this explanation, "Why are pill bottles created the way that they are? Typical answers include fitting on the pharmacy shelf, fitting on the existing packaging line, meeting guidelines for child safety, etc… Most of today's pill bottles are all the same. They have cotton in them and an annoying foil wrapper. Pharmacists pour the drug out, count the tablets and transfer them to plain brown bottles with typed labels. Why do half the bottles out there still not have the ability to put a prescription label directly on them? Many of the bottles are small because the pills are small, yet we miss the opportunity to make them large enough to be more practical because we want to keep our eco-footprint small. So we make the bottles smaller so that we as pharmaceutical companies can say, 'Well we're cutting our use of materials by X percent.' Have we really saved the environment when the pharmacist uses yet another brown bottle that does not indicate who made the medication inside, does not have a lot number or any of the things that might provide value to a consumer. This value is lost in the transfer from manufacturer pill bottle to pharmacy pill bottle."

REAL-WORLD EVIDENCE

Payers are increasingly insisting that pharmaceutical brands supply real-world evidence and data on product efficacy and cost savings. Given the adherence issues, real-world data is probably more accurate than clinical trial data that was gathered under perfect conditions, using a relatively small patient population, and often conducted years ago for older brands. Real-world data, more easily accessible now than in the past, is starting to shine a light on what pharmaceutical products are really doing versus what they did in the clinical trial setting.

"Digital registry studies are a real-world experience that goes beyond the often-criticized perfect environment of clinical trials. They provide an opportunity for us to go to the payers with real-world evidence, budget impact analysis, and budget impact modeling," says Peter Justason.

Payers are insisting on real-world evidence that a particular drug therapy is the best option.

"Brands can start to offer wearable technology for a disease state or offer other digital support services that help build patient adherence or persistence and demonstrate actual outcomes through EMRs where data capture is happening as never before. We are starting to quantify to what extent in the real world a treatment with a given product actually lowers overall healthcare costs. If we can aggregate that data across health plans, we can demonstrate to Medicare why it should cover the brand. Going forward," says Lars Merk, "this model is really not optional… It is the model."

PERSONALIZED MEDICINES

"Pharmaceutical products are effective, on average, about 60 percent of the time. So for about a third of all people who take a particular medicine, it doesn't work. This is why we have second-line and third-line therapies. In pharma, we won't move to a winner-takes-all scenario, even under managed care. But we might do so in the medical device industry because doctors and patients need options," says Lars Merk.

"With the mapping of the human genome and other advances, personalized medicine is becoming more possible and affordable. However, society will have some tough choices in this area as well when it comes to privacy. Will patients trade their genome information for a better medicine tailored to their particular DNA or body chemistry? How will this data be shared or used by the insurance company paying for the medicine? Think how 23andMe could change the physician-patient history conversation. Today a doctor asks whether anybody in the family has a history of stroke, heart attack, and so on. Many patients simply don't know or remember their family history, or it was simply never shared with them. Medical files rely on often-inaccurate self-reported data. As a result, we often don't even know where we should start intervening. If we were to intervene in the case of a 20-year-old who had diabetes or cardio-vascular concerns due to that patient's genome," Lars Merk says, "we might get a change in behavior earlier rather than starting the conversation at age 50. Now that would affect a positive health outcome."

From Scott Richardson's perspective, we need to do a better job of understanding our customer to deliver personalized healthcare in a meaningful way to patients. He draws the analogy from Netflix and

Amazon. "When I go on Amazon and I look for something," he says, "they indicate people who bought that book often also buy these books. Netflix tells you that people who liked that movie also liked these movies and actually suggest other titles. In medicine, people with high blood pressure often have a cholesterol issue, and people on pain medications are often depressed. Patients are more than just a singular condition."

PERSONALIZED MARKETING

Brand messaging and adherence programs today tend to be one-size-fits-all. In reality, each individual will need different motivators and services. Some may need information; for others, the issue is financial; and others need social stimulus. Adherence and educational programs need to be as flexible and as personalized as the medicines we are striving to produce.

Pharmaceutical marketers tend to create one ad and serve it up to everyone to save money. The right approach is having 40 ads that target the 40 different needs of physicians or consumers at the right place, at the right time, in the right location. It's not enough for our companies to demand targeted advertising without actually doing something useful and valuable with that data, Merk says. This is a discipline where pharmaceutical marketers need training.

Why build a separate adherence app that reminds people to take their medicine rather than trying to better serve patients by fitting our product more conveniently into their lives? For example, why not, instead, work with Microsoft to build a plug-in for Outlook or figure out a way to integrate with Evernote? We don't create magazines and conferences; we message through the ones that customers are already using.

"The fundamentals of marketing don't change, but the tools we use are changing. To be effective," Merk says, "marketers of the future need to understand the tools available to them but still retain the marketing principles they learned so many years ago."

GOING BEYOND WEARABLES

"Wearables today face a compliance and adoption challenge. The majority of wearables are used for too short a period of time. There is a generation of 'transparent wearables' starting to emerge where little or no programming is required. Just put it on. It generates the data and communicates it to where it needs to go. For example, this data may go to the patient, with a summary to her physician, and the wearable may present the right green, yellow, or red flags at the appropriate time providing encouragement, necessary warnings, or alerts to the most appropriate party. Mobile has some of this benefit today in that data flows to a back-end system to prove an action's been taken. If we combine that information with an electronic medical or health record, we can correlate actions with outcomes. The effective use of mobile coupled with wearable devices, tied in to big data, and serving up an adherence program through, say, gamification might just prove outcomes and meet the new emerging reimbursement standard," says Craig DeLarge.

"Imagine a pill with a biosensor in it such that it will send a signal to your iPhone, which then can send it to your electronic medical record, the insurance company, the doctor's office, the pharma company, and so on, and will tell if you took your pill so that it can directly track compliance and adherence. Over time, as the pill dissolves, you will continue to receive signals from your system about how you are reacting to the pill. The biosensors will also report in on safety and efficacy measures on an individual patient level. This is the category of 'ingestibles,'" says Craig DeLarge.

While costs are still a challenge, this technology exists today. Pharmaceutical companies are already combining diagnostics with a pharmaceutical, software, and sensors in partnership with companies like Proteus. Proteus is leveraging microchip technology small enough to sit on a pill. It sends the signal to a patch worn on the skin, which then communicates with the patient's smartphone. Bundling multiple solutions into a single offering leads to a more robust discussion with payers and a better means to measure outcomes.

BIG DATA

"Pharma has a lot of data, but it often exists in silos that are used in relatively mis- and disintegrated and unsophisticated ways. I think the industry tends to confuse data volume with data sophistication," says DeLarge.

Merk says, "We are nearing a point where it won't matter if people are on their phone, their tablet, or standing in front of their open refrigerator. No matter what they are doing, they will be influencing and adding data to their personalized data history."

Pharma's hands are still tied today because we only have the ability to look at limited data sets. The data are not all being aggregated. Companies such as Flatiron Health are doing some interesting things, tying together oncology data from cancer centers all over the country, Merk says. "This is probably one of the most innovative data sets around. We help maybe 4 percent of patients who get involved in clinical trials, but this data could help the other 96 percent. None of this is without challenges, due to HIPAA and other regulations, but people eventually are going to be a little bit freer with their data. HIPAA from a privacy standpoint does need to evolve. And I think that everyone is recognizing that. I think there will be a big push for changes to HIPAA because the greatest beneficiary of a change is actually the federal government."

"In managing the health economy, we should be trying to help understand and improve on real-world data," says Merk. "This is where outcomes research comes in. It's not just about how much the cholesterol was lowered in the clinical trials. It's about how many heart attacks these patients have. How long do they live? What's their quality of life?"

PRIVACY

Data that others entrust to us is another risk, or opportunity, for pharma, depending on how we choose to handle it. Data is power. But power can corrupt. Our path to both opportunities and risks is being paved today by the choices we make in data stewardship.

It is worth remembering that the historic gifts to physicians and fancy dinner meetings all led to the Sunshine Act. We have, historically, failed to self-police our own efforts strongly enough. How are you stewarding data today? Are you providing valuable customized services, or are you bombarding your customers with things that probably don't provide value?

"In general, physicians don't mind us collecting data about them if we are using that data to provide better value and outcomes," says Lars Merk.

ORGANIZATIONAL STRUCTURE

Most executives we interviewed indicate that an organizational overhaul is required to truly embrace and capitalize on the changes in front of us. The industry, to some extent, is trapped in Clayton Christensen's *Innovator's Dilemma,* having long benefited from very high margins and patent-protected drugs. It can be hard to gain funding for an initiative that does not deliver the same margin as the current portfolio of brands, even when that initiative may be the foundation of the organization's go-forward success in a changing landscape.

Successful organizations are the ones that understand, at the senior level, that a change management skill set is required and they must modify the incentive structure accordingly.

"We lived through the center-of-excellence teams, going back to the '90s when the Internet first started to infiltrate healthcare marketing," Merk says. "They are back again to help cope with new changing times, and that's a good thing. These teams should have a limited lifespan in which, if they do their jobs, they successfully educate and lead change and then put themselves and their center out of business. We need more disruptive people in the industry as opposed to just disruptive technologies; we need people who are willing to be comfortable with the ambiguity and the pace of change."

Looking at things from the customer's point of view radically changes our behavior and requires that not only should the marketing team be customer centric but also every other component of our infrastructure, including med/legal, research, manufacturing, and so on must also be customer-centric.

Is your company ready for the changes ahead? Are you ready to deliver results? They are the future of pharmaceutical and healthcare marketing.

CONCLUSION

The future of pharmaceutical and healthcare marketing can be summed up in the single word that is the title of this book: RESULTS.

If pharmaceutical and other life sciences companies hope to market their products in a way that will remain competitive, they must adapt to the demands of the new era of healthcare. Through the chapters of this book, we have focused on the trends that you need to understand as the perfect storm brews. This is the knowledge you need to survive. Without it, you will be left behind as other industries disrupt healthcare.

When we discussed regional and digital marketing and the rapidly evolving systems of healthcare in the era of big data, we emphasized above all that we all are in the business of serving people. Those people are often in dire need of good advice and reliable information, and the industry today has the potential to provide both better than ever before.

The informational and technological revolutions have forever changed the practice of medicine. We can glean and analyze relevant data in a flash, and marketers can deliver it with pinpoint accuracy at just the right time.

We have an unprecedented opportunity to inform and to serve, and in doing so, we can thrive as an industry. So long as we remember that we are serving real people who are often going through the most difficult battles of their lives, we will do well by doing good. Customers need personal, meaningful, and practical communication, and when they get it, they respond favorably to whoever serves it up in their preferred format at the right time.

Pharma companies today need to add value beyond the pill by adapting, serving, and collaborating with the other players in the healthcare universe. They need to show that they can deliver results better than anyone else can. That's how the industry will not just remain relevant but lead in the new era of outcomes-based healthcare.